T0341323

Optimal Power Flow
Using FACTS Devices

Optimal Power Flow Using FACTS Devices

Soft Computing Techniques

L. Ashok Kumar and K. Mohana Sundaram

CRC Press is an imprint of the
Taylor & Francis Group, an **informa** business

MATLAB® is a registered trademark of The MathWorks, Inc. For product information, please contact: The MathWorks, Inc. 3 Apple Hill Drive Natick, MA 01760-2098 USA Tel: 508-647-7000 Fax: 508-647-7001 Email: info@mathworks.com Web: www.mathworks.com

First edition published 2021

by CRC Press
6000 Broken Sound Parkway NW, Suite 300, Boca Raton, FL 33487-2742

and by CRC Press
2 Park Square, Milton Park, Abingdon, Oxon, OX14 4RN

© 2021 Taylor & Francis Group, LLC

CRC Press is an imprint of Taylor & Francis Group, LLC

ISBN: 978-0-367-56572-5 (hbk)
ISBN: 978-1-003-09842-3 (ebk)

Typeset in Times LT Std
by KnowledgeWorks Global Ltd.

Contents

Preface

The electrical power service providers are subjected to many constraints such as reliability, quality, and minimizing generating cost and emission for the supply of electric power to its consumers. Another major problem is to transmit the power to remote areas. The application of series compensation increases the power transfer capability of the line, but with introduction of subsynchronous resonance (SSR) oscillations. The invention of flexible alternating current transmission system (FACTS) devices ensures optimal generating cost, emission, power loss with improved voltage, and also use of SSR oscillations for mitigation in electrical power networks under different levels of series compensation. The application of soft computing techniques concept of optimization finds prominent place in providing optimal power flow (OPF). This book not only covers the entire scope of OPF using soft computing techniques but also explains the application of FACTS devices with intelligent controllers for damping subsynchronous resonance oscillations. It is well supported by the extensive literature survey, algorithms to solve OPF, and case studies. The book will certainly be useful to undergraduate and postgraduate students as well as to academicians, power engineers, researchers, scientists, and practitioners.

L. Ashok Kumar
K. Mohana Sundaram

Acknowledgments

The authors are always thankful to the Almighty for His perseverance and achievements. The authors owe their gratitude to Shri L. Gopalakrishnan, Managing Trustee, PSG Institutions; Shri. K.P. Ramasamy, Chairman, KPR Institute of Engineering and Technology, Coimbatore, India; the authors also owe gratitude to Dr. K. Prakasan, Principal In-Charge, PSG College of Technology, Coimbatore, India; and Dr. M. Akila, Principal, KPR Institute of Engineering and Technology, Coimbatore, India, for their wholehearted cooperation and great encouragement in this successful endeavor.

L. Ashok Kumar would like to take this opportunity to acknowledge those people who helped him in completing this book. "I am thankful to all my research scholars and students who are doing their project and research work with me. But the writing of this book is possible mainly because of the support of my family members, parents, and sisters. Most importantly, I am very grateful to my wife, Y. Uma Maheswari, for her constant support during writing. Without her, this book would not have been possible. I would like to express my special gratitude to my daughter, A.K. Sangamithra, for her smiling face and support; it helped a lot in completing this work. I would like to dedicate this work to her."

K. Mohana Sundaram would like to take this opportunity to acknowledge those people who helped him in completing this book. "I am thankful to all my research scholars who are doing their project and research work with me. But this book has been possible mainly because of the support of my family members. I am very grateful to my wife, R. Nidhya, for her constant support during writing. I would like to express my special gratitude to my daughters M.N. Soumitra and M.N. Muhilsai for their support; it helped a lot in completing this work."

About the Authors

L. Ashok Kumar was a postdoctoral research fellow from San Diego State University, California. He is a recipient of the BHAVAN fellowship from the Indo-US Science and Technology Forum and SYST Fellowship from DST, Government of India. His current research focuses on integration of renewable energy systems into smart grid and wearable electronics. He has 3 years of industrial experience and 19 years of academic and research experience. He has published 167 technical papers in international and national journals and has presented 157 papers in national and international conferences. He has completed 26 Government of India funded projects, and currently 7 projects are in progress. His Ph.D. work on wearable electronics earned him a National Award from ISTE, and he has received 24 awards on the national level. Ashok Kumar has seven patents to his credit. He has guided 127 graduate and postgraduate projects. He has also produced 4 Ph.D. scholars and 12 scholars are pursuing their Ph.D. work. He holds membership as well as prestigious positions in various national forums. He has visited many countries for institute industry collaboration and as a keynote speaker. He has been an invited speaker in 178 programs. Also, he has organized 72 events, including conferences, workshops, and seminars. He completed his graduation in electrical and electronics engineering from University of Madras, postgraduation from PSG College of Technology, and Master's in Business Administration from IGNOU, New Delhi, India. After his graduation, he joined as project engineer at Serval Paper Boards Ltd., Coimbatore (now ITC Unit, Kovai). At present, he is working as Professor and Associate HoD in the Department of EEE, PSG College of Technology, and is also doing research work in wearable electronics, smart grid, solar PV, and wind energy systems. He is also a Certified Charted Engineer and BSI Certified ISO 500001 2008 Lead Auditor. Dr Kumar has authored the following books in his areas of interest: *Computational Intelligence Paradigms for Optimization Problems Using MATLAB®/SIMULINK®,* CRC Press; *Solar PV and Wind Energy Conversion Systems—An Introduction to Theory, Modeling with MATLAB/SIMULINK, and the Role of Soft Computing Techniques,* Green Energy and Technology, Springer; *Electronics in Textiles and Clothing: Design, Products and Applications,* CRC Press; *Power Electronics with MATLAB,* Cambridge University Press; *Automation in Textile Machinery: Instrumentation and Control System Design Principles,* CRC Press, Taylor & Francis Group; *Proceedings of International Conference on Artificial Intelligence, Smart Grid and Smart City Applications,* Springer International Publishing; *Deep Learning Using Python,* Wiley India Publications; *Monograph on*

Smart Textiles; *Monograph on Information Technology for Textiles*; and *Monograph on Instrumentation & Textile Control Engineering.* He is Senior Member in IEEE and Fellow Member in IE (India), IETE, and IET (UK).

K. Mohana Sundaram is a Professor in Department of EEE at KPR Institute of Engineering and Technology, Coimbatore India. He has 18 years of teaching and research experience. His current research interests include intelligent controllers, power systems, embedded systems, and power electronics. He has completed a funded project of worth Rs. 30.84 lakhs sponsored by DST, Government of India. He received his B.E. degree in Electrical and Electronics Engineering from University of Madras in 2000, M. Tech degree in High Voltage Engineering from SASTRA University in 2002, and Ph.D. degree from Anna University, India, in 2014. Under his supervision, four candidates have completed their Ph.D. from Anna University, Chennai, while nine candidates are still pursuing. He has published 47 articles in international journals. Prof. Sundaram serves as reviewer for IEEE journals, Springer journals, and Elsevier. Prof. Sundaram is a member of IE, ISTE, IAENG.

1 Introduction

LEARNING OUTCOME

 i. To understand the basic terminologies behind power system network.
 ii. To learn about the power system stability.
 iii. To study about the power system deregulation.
 iv. To study about the FACTS controllers and their types.

1.1 AN OVERVIEW OF POWER SYSTEMS

The electrical system consists of the generation, transmission, and distribution as shown in Figure 1.1. Electricity generation is the conversion of one form of energy into electricity. Electricity is generated from nonrenewable sources such as hydro, thermal, and nuclear power stations and from renewable sources such as solar, wind, etc. The voltage generated is usually 6.6 kV, 10.5kV, 11 kV, 13.8 kV, 15.75 kV, etc. The bulk power is transmitted through the transmission system and power to the consumers through distribution systems.

1.1.1 ELECTRICAL SYSTEM COMPONENTS

The various features of the electrical system are shown in Figure 1.2 and are discussed below.

Generators: Convert mechanical energy into electricity power
Transformer: Step-up/step-down voltage from one end to another without change in frequency
Transmission lines: Transmits power from one end to other
Control equipment: To provide security to all operating equipment
Primary transmission: 110 kV, 132 kV, or 220 kV or 400 kV or 765 kV transmission line by 3Ø 3-wire system
Secondary transmission: 3Ø 3-wire system, using either 33 kV or 66kV feed
Primary distribution: 3Ø 3-wire system, 11 kV or 6.6 kV
Secondary distribution: 3Ø400 V, 1Ø230 V

1.1.2 GENERATORS

A generator converts mechanical energy into electrical energy. Generating voltages are 6.6 kV, 10.5 kV, or 11 kV, which can increase up to 110 kV/132 kV/220 kV at the generating side to reduce current in the transmission line and to reduce transmission losses. Generators generate real power (MV) and reactive power (MVAr).

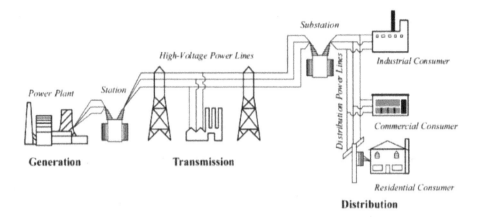

FIGURE 1.1   Electric power flow layout.

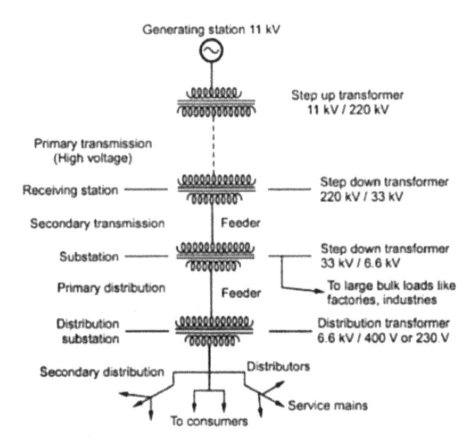

FIGURE 1.2   Structure of power system.

1.1.3 Transformers

It is a fixed instrument that transfers power from one region to another with constant frequency. The main function of transformers is to increase power from low generation levels to high generation levels and also decrease explosion rates from high power transmission rates to low transmission rates. When the transmission voltage is increased, the current flow of the grid decreases; thus the transmission loss (I^2R) decreases.

1.1.4 Control Equipment

Circuit Breaker is used to turn on and off the circuit under normal and abnormal conditions (fault). Various breakers are oil, air blast, vacuum, and SF6 circuit breakers. During the transfer the error status will enable the circuit breaker to operate.

Isolators are placed in substations to isolate the part of system during maintenance. It can operate only during no-load condition. Isolated switches are provided on each side of the circuit breaker.

Busbar connects various lines operating at the same voltage electrically. It is made of copper or aluminum. Various types of bus arrangements include single busbar, double busbar, ring busbar, etc.

1.1.5 Transmission System

It supplies only large blocks of power to bulk power stations or very big consumers. It interconnects the neighboring generating stations into a power pool, i.e., the interconnection of two or more generating stations. Tolerance of transmission line voltage is ±5% to ±10% due to the variation of loads.

1.1.6 Primary Transmission

The generated power is then transmitted through transmission lines. The voltage is stepped-up and the line current is reduced to reduce the power loss. Generating stations are installed with step-up transformers to boost the voltages to higher values. Power is transmitted to the receiving end substations from sending end substations through high-voltage 3-phase 3-wire lines. Primary transmission voltages are 110 kV, 132 kV, 220 kV, 400 kV, or 760 kV.

1.1.7 Secondary Transmission

At the receiving end input, the voltage is reduced to a value of 66 kV or 33 kV or 11 kV using step-down transformer. The secondary transmission line creates a connection between the receiving end and secondary station. It uses a 3-phase 3-wire system, and the operators it uses are called feeders.

1.1.8 DISTRIBUTION SYSTEM

The part of the power supply that connects all consumers to a location in a power source in bulk or a transmission line is called a distribution system. The distribution channel spreads power to domestic, commercial, and small consumers. Distribution transformers are usually installed on stables or on a mounted plinth at or near consumers.

1.1.9 PRIMARY DISTRIBUTION

At the secondary substations, the voltage is reduced to 11 kV or 6.6 kV using step-down transformers. The primary distributor forms the link between the secondary substation and distribution substation, and the power is fed into the primary distribution system. It uses 3-phase 3-wire system.

1.1.10 SECONDARY DISTRIBUTION

The voltage is stepped down to 400 V (3 phase) or 230 V (1 phase) using step-down transformers. The distribution lines are drawn along the roads and service connections to the consumers are tapped off from the distributors and use 3-phase 4-wire system. Single-phase loads are connected between a phase wire and neutral wire.

1.2 POWER SYSTEM STABILITY

The power system comprises certain machines that operate in synchronism. In order to continue energy system flow, synchronization has to be maintained in all solid-state conditions. Due to the disturbance occurring in a system, energy is developed because it becomes normal or stable.

The capacity of the system to get back to its normal or stable state after disturbance is called stability. Disturbance of the system can occur due to sudden load change, short circuits like LG, LL, 3Ø faults, etc.

The system stability depends on the behavior of the synchronization equipment following the interruption. The stability limit is defined as the maximum power to flow through a particular part of the system under line disturbances or fault. The power system stability is divided into two types as shown in Figure 1.3.

- Steady-state stability
- Transient stability

Steady-state stability: The capability of the system to restore its synchronization (speed and frequency of all the networks are the same) after a slow and small disturbance due to gradual power change is called steady-state stability. Steady-state stability is divided into two parts

- Dynamic stability
- Static stability

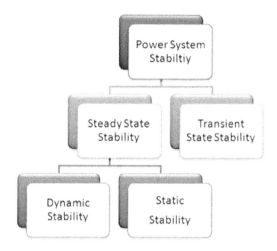

FIGURE 1.3   Power system stability classification.

Dynamic stability – Indicates the stability of the system to reach its stable state after a very small disturbance (disturbance occurs in 10–30 seconds). It is also known as low signal intensity. It is mainly due to fluctuations in load or generation rate.

Static stability – Means the stability of an available system without the help (benefit) of automatic control devices such as controllers

Transient stability: It is defined as the ability of the system to return to its normal state after a major disturbance. Severe disruptions occur in the system due to sudden removal of the load, line shifting operations, fault, line outage, etc.

Transient stability is conducted for planning of new transmitting and generating systems. The swing equation describes the performance of the synchronous machine during disturbances.

The transient and steady-state disturbances occurring in the power system are shown in Figure 1.4. These disturbances reduce machine synchronization and the system becomes unstable.

Stability studies are useful in determining the critical clearing time of breakers, voltage levels, and transmission capacity of the power systems.

1.3   POWER SYSTEM DEREGULATION

A **vertically integrated company** regulated by state authority includes generation subsystem, transmission subsystems, and distribution subsystems that were responsible for serving all its customers. Prior to 1992, within a specific area, one company owned and operated all employees under the program.

Customers were captured at a local company, and electricity prices were set by the regulating business in each state. Prices include the cost of generating, operating, and maintaining the transmission and distribution systems included in the customer pricing in this area; large customers had some bargaining ability, but small- and medium-sized customers were left with unattractive options at low-cost management centers.

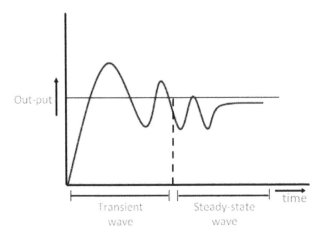

FIGURE 1.4   Steady-state and transient waveform.

To facilitate the growth of the market's electricity capacity and to promote competition, the **National Energy Policy Act (NEPA)** encourages

- **Integrated resource management (IRP) and demand-side management (DSM)**
- **Exempted wholesale generators (EWGs)**
- Resources that are permitted to purchase from the corresponding **EWG** under the jurisdiction of the state commission

The **Federal Energy Regulatory Commission (FERC)** has been authorized to force contracts to be signed as the contract is for the public benefit. This is the first step to open transfer access.

1.3.1 Deregulation

An electric industry deregulation **Notice of Proposed Rulemaking (NOPR)** was issued by **FERC** in 1993. The vertically integrated structure was unbundled by NOPR into

- **GENCO (generation company)**, which provides electric generation and maintains contracts with ancillary services to improve reliability and power quality to customer.
- **TRANSCO (transmission company)**, which transfers power from generation system to distribution system through the transmission grid with the support of ancillary services.
- **DISCO (distribution company)**, which provides physical connection for end users or customers.

The three companies are horizontally integrated. They are maintained by a central coordinator or **independent contract administrator (ICA)**. The **ICA** acts like an

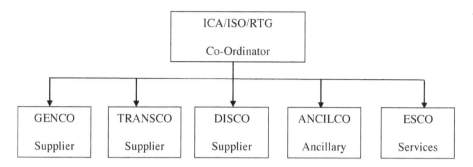

FIGURE 1.5   Electric utility industry structure in deregulated environment.

agent between generations and end users. It handles sale and purchase of power and rewards the benefit of profits.

The ICA can be started as two separate entities, which are the exchange of power (which is a competitive power market) and the **independent system operator (ISO)**.

The ancillary service will be available by ISO. Ancillary services may be provided by a third party, called as **ancillary service company (ANCILCO)**. Emerging players in the new area are **energy services companies (ESCOs)**. This could be a large industrial customer or a customer stake. These companies purchase power from GENCOs to meet the customers' needs. ESCO will also partner with DISCOs to find a way to distribute to their customers. Figure 1.5 shows this outline.

1.3.2 ANCILLARY SERVICE

A large number of functions in a connected grid are referred to as ancillary services. The **North American Electric Reliability Council (NERC)** and the **Electric Power Research Institute (EPRI)** have identified 12 ancillary services:

1. **Regulation:**
   Regulate generation to load balance in a control area.
2. **Load following:**
   Maintain load-generation balance till the scheduling period.
3. **Energy imbalance:**
   Use generators to meet hourly and daily variations in loading.
4. **Operating reserve (spinning):**
   The unloaded generator is synchronized to the grid to respond to generation-load imbalances due to any outages. It is available at all times.
5. **Operating reserve (supplemental):**
   The unloaded generator serves the same as the spinning reserve but does not necessarily respond immediately to load generations caused by any outages. It is available at all times.
6. **Backup supply:**
   Generating stations guarantee supply to consumers both in scheduled and unscheduled situations.

7. **System control:**
   Human brain function is compared with this activity. Its main function is to maintain generation-load balance in scheduled and unscheduled operator functions.

8. **Dynamic scheduling:**
   Hardware control includes computers interfaced with real-time meters and telemeters.

9. **Reactive power and voltage control support:**
   The absorption and injection of reactive power to sustain system voltages within the limit.

10. **Real power transmission losses:**
    These balance the diversity between energy from the generator and energy to the consumer.

11. **Network stability services from generation sources:**
    Special equipment like PSS, dynamic braking, etc., is used to maintain secured transmission function.

12. **System black start capability:**
    The capability of the generating unit to restart after shutdown on its own and support other grid-connected units to restart after blackout.

1.3.3 BILATERAL TRANSACTIONS

All the vertically integrated utilities are transformed into ISOs such as GENCOs, TRANSCOs, and DISCOs.

Bilateral transaction is that where a consumer can have a contract to get power from any GENCO through any TRANSCO via any DISCO from one or any other area.

1.3.4 DISCO PARTICIPATION MATRIX

The contracts between DISCOs and GENCOs form the **DISCO Participation Matrix (DPM)**. The rows are GENCOs and columns are DISCOs participating in the contract, for a two-area system, DPM will be

$$
\begin{array}{c}
 \\
\text{AREA 1} \\
\text{AREA 2}
\end{array}
\begin{array}{c}
 \\
1 \\
2 \\
3 \\
4
\end{array}
\overset{\begin{array}{cccc} 1 & 2 & 3 & 4 \end{array}}{
\left(
\begin{array}{cccc}
cpf_{11} & cpf_{12} & cpf_{13} & cpf_{14} \\
cpf_{21} & cpf_{22} & cpf_{23} & cpf_{24} \\
cpf_{31} & cpf_{32} & cpf_{33} & cpf_{34} \\
cpf_{41} & cpf_{42} & cpf_{43} & cpf_{44}
\end{array}
\right)
}^{\text{DISCO} \rightarrow}
\quad \text{GENCO} \downarrow
$$

where cpf_{jd} = contract participation factor of jth GENCO in the load following dth DISCO.

1.4 FACTS DEVICES

The ideology of FACTS was first developed by N. G. Higorani in 1986, which transformed traditional methods of transmission into modern technology.

FACTS is defined by the IEEE as "a power electronics based system and other static equipment that provide control of one or more AC transmission system parameters to enhance controllability and increase power transfer capability."

The design of FACTS devices comprises of integration of traditional components like transformers, reactors, switches, and capacitors with power electronics elements like thyristors and transistors.

FACTS devices made of static power-electronic devices are placed in AC transmission networks for the increase in power transfer capability, stability, and controllability. They are installed for either series and/or shunt compensation and/or hybrid. They also serve for congestion management and loss optimization.

FACTS devices are used to increase the capability of existing and new transmission and distribution networks. The benefits of FACTS device installation are

- Steady-state and dynamic reactive power compensation and stability enhancement
- Voltage regulation
- Power quality improvement
- Power transfer capability improvement
- System reliability and operation flexibility improvement
- Fault current reduction
- Transmission losses reduction

FACTS devices are simple energy conversion devices that control the transmission power flow with the usage of high-power semiconductor switches developed for large current electronic switching and high-voltage switching.

The FACTS device core is developed by the integration of manufacturing technology, modern control technology, and traditional power grid technology in electric power system. The traditional mechanical breaker is replaced with electronic circuits.

1.4.1 LOADING CAPABILITY LIMITS

The FACTS for the transmission system is shown in Figure 1.6. The active power transmission shall be done till it reaches the thermal limit, whereas the voltage and stability limits can be adjusted with the usage of FACTS devices.

FACTS devices are much needed with increase in line length. The impact of FACTS devices is attained through switching power electronic devices that work in very short time—to less than a second.

1.4.2 THERMAL LIMIT

Concerning overhead transmission lines, temperature, wind conditions, conductor conditions, and ground clearance are the functions of thermal capacity.

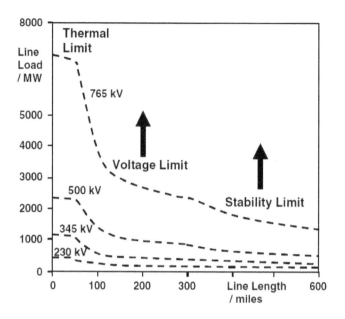

FIGURE 1.6   Operating limits of transmission lines for different voltage levels.

1.4.3   DIELECTRIC LIMIT

The operating voltages of the lines can be increased by 10% or more. FACTs devices ensure safe operation under permissible overvoltage conditions.

1.4.4   STABILITY LIMIT

The FACTs devices can overcome transient stability, dynamic stability, steady-state stability, frequency, and voltage collapse.

1.5   TYPES OF FACTS CONTROLLERS

FACTs devices made of power electronic controllers are used to improve system controllability and power transfer capability in AC systems. The FACTS controllers are

1. Series-connected controllers
2. Shunt-connected controllers
3. Combined series-series controllers
4. Combined shunt-series controllers

1.5.1   SERIES CONTROLLERS

Voltage is injected in series with the line in series controllers as shown in Figure 1.7. It generates or absorbs reactive power when voltage and current are in phase quadrature.

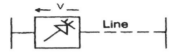

FIGURE 1.7   Schematic diagram of series controller.

Series controllers are

   i. Static synchronous series compensator (SSSC)
  ii. Thyristor-controlled series compensation
 iii. Thyristor switched series capacitor (TSSC)
  iv. Thyristor switched reactor (TSR)
   v. Thyristor-controlled series capacitor (TCSC)
  vi. Thyristor-controlled series reactor (TCSR)

1.5.2   SHUNT CONTROLLERS

Currents are injected by the shunt controllers in the connection point as shown in Figure 1.8. It generates or absorbs reactive power when voltage and current are in phase quadrature.
   The shunt controllers are

   i. Static synchronous compensator (SSC)
  ii. Static VAR system (SVS)
 iii. Static VAR compensator (SVC)
  iv. Static condenser
   v. Thyristor-controlled reactor (TCR)
  vi. Thyristor-switched capacitor (TSC)
 vii. Thyristor-switched reactor (TSR)

1.5.3   COMBINED SERIES-SERIES CONTROLLERS

Different series controllers are integrated and controlled in the order as shown in Figure 1.9. It delivers real power and self-regulating series reactive power compensation. The interline power flow controller (IPFC) is an example of a series-series controller.

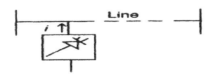

FIGURE 1.8   Schematic diagram of shunt controller.

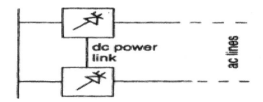

FIGURE 1.9   Schematic diagram of series-series controller.

1.5.4   COMBINED SERIES-SHUNT CONTROLLERS

Different series controls and shunt controls are integrated and operate in an integrated manner as shown in Figure 1.10.

The different combined series shunt controllers are

    i. Unified power flow controller (UPFC)
    ii. Interphase controller (IPC)
    iii. Thyristor-controlled phase shifting transformer (TCPST)

1.5.5   CLASSIFICATION BASED ON DEVICES USED

Depending on the power electronic devices that are used in control, the classification of FACTS devices are

    i. Variable impedance type
    ii. Voltage source converter (VSC) based

The variable impedance types are

    i. SVC
    ii. Thyristor-controlled series capacitor (TCSC)
    iii. TCPST

The VSC types are

    i. Static synchronous compensator (STATCOM)
    ii. Static synchronous series compensator (SSSC)

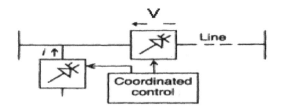

FIGURE 1.10   Schematic diagram of series-shunt controller.

iii. Interline power flow controller (IPFC)
iv. UPFC

1.5.6 SPECIAL PURPOSE FACTS CONTROLLERS

The FACTS controllers used for special purposes are

i. Thyristor-controlled braking resistor (TCBR)
ii. Thyristor-controlled voltage limiter (TCVL)
iii. Thyristor-controlled voltage regulator (TCVR)
iv. Interphase power controller (IPC)
v. NGH-SSR damping

1.5.7 FACTS CONTROLLERS BENEFITS ARE

The FACTS controller benefits are

i. Voltage support at the critical buses is provided
ii. Voltage is improved
iii. Power loss is reduced
iv. Control of power flow is ordered
v. Lines are upgraded
vi. High system security
vii. Suitable for interconnection of power grids of different frequencies
viii. Transient stability is increased

As the transient stability increases, the dynamic system security improves and thereby blackouts are reduced.

i. Steady-state stability is increased
ii. Voltage fluctuations are avoided

1.5.8 APPLICATIONS OF FACTS CONTROLLERS

The major application of FACTS controllers is to improve the power quality in distributed systems.
  The various applications are

i. Power flow control
ii. Increase the power transfer capability
iii. Voltage control
iv. Power conditioning
v. Flicker mitigation
vi. Interconnection of renewable energy generation system and distributed energy
vii. Generation
viii. Stability improvement

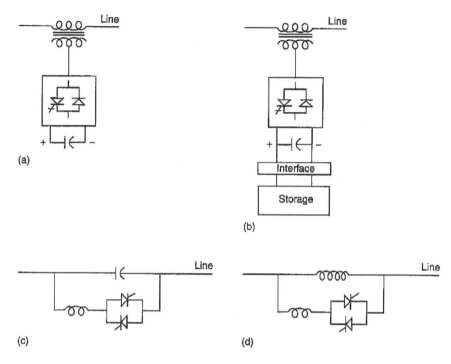

FIGURE 1.11 Series controllers: (a) static synchronous series compensator (SSSC); (b) SSSC with storage; (c) thyristor-controlled series capacitor (TCSC) and thyristor switched series capacitor (TSSC); (d) thyristor-controlled series reactor (TCSR) and thyristor switched reactor (TSSR).

1.6   SERIES CONTROLLERS

A series controller be variable impedance like capacitor, reactor, or variable source like frequency, subsynchronous frequencies to meet the need.

- Inject voltage in series with the line.
- Absorbs or injects reactive power by the series controllers as the voltage and current are in phase quadrature.

Some of the series controllers, SSSC, TCSC, and TCSR, are shown in Figure 1.11. They can be used in the transmission system for power flow control and damp oscillation.

1.7   SHUNT CONTROLLERS

- A variable impedance or variable source or combination of both can be a shunt controller.
- Current is injected at the connection point by the shunt controllers.
- Absorbs or injects reactive power by the shunt controllers as the voltage and current are in phase quadrature.

Some of the shunt controllers like static synchronous generator (SSG) and SVC are shown in Figure 1.12.

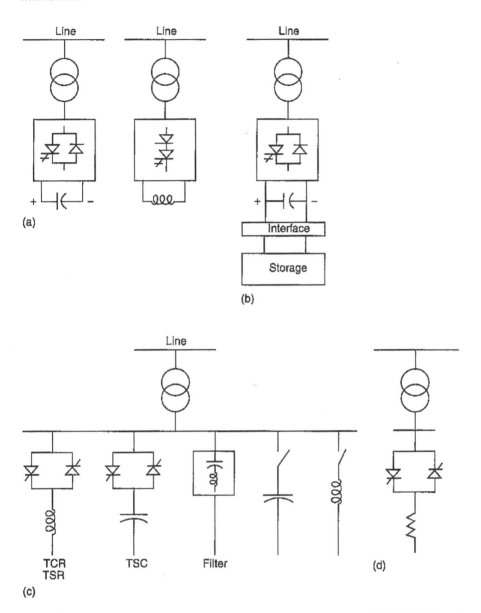

FIGURE 1.12 Shunt controllers: (a) static synchronous compensator (STATCOM) based on voltage-sourced and current-sourced converters; (b) STATCOM with storage, i.e., battery energy storage system (BESS) superconducting magnet energy storage and large DC capacitor; (c) static VAR compensator (SVC), static VAR generator (SVG), static VAR system (SVS), thyristor-controlled reactor (TCR), thyristor-switched capacitor (TSC), and thyristor-switched reactor (TSR); (d) thyristor-controlled braking resistor.

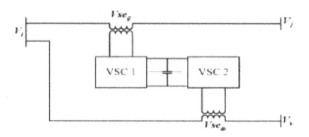

FIGURE 1.13   Interline power flow controller.

1.8   COMBINED SERIES-SERIES CONTROLLERS

- This can be a combination of different series controllers in a multiple line system.
- Transfer real power among line and reactive power compensation for each line.
- Balance both real and reactive power flow.
- The IPFC shown in Figure 1.13 is an example of a combined series-series controller. In UPFC, the support system is a shunt converter, and in IPFC it is a series converter.

1.9   COMBINED SERIES-SHUNT CONTROLLERS

- A combination of series and shunt controllers in a coordinated manner. UPFC is an example of combined series-shunt controllers.
- Current is injected by the shunt part of the controller, and voltage is injected by the series part of the controller.
- A unified series and shunt controller supports in sharing real and reactive power between the series- and shunt-connected DC power link.

UPFC and thyristor-controlled phase-shifting transformer (TCPST) shown in Figure 1.14 are few examples.

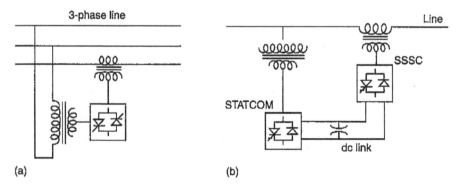

FIGURE 1.14   Combined series-shunt controllers: (a) thyristor-controlled phase shifting transformer (TCPST) or thyristor-controlled phase angle regulator (TCPR); (b) unified power flow controller (UPFC).

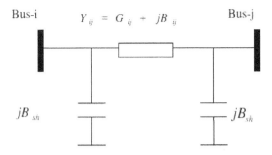

FIGURE 1.15 Model of transmission line.

1.10 THYRISTOR-CONTROLLED SERIES CAPACITOR (TCSC)

Thyristor-controlled series compensator (TCSC) is a series connected in the transmission line.

- The controllable and equivalent reactance can be inserted and connected in a series with a line for compensation of line inductance.
- Therefore, the total transfer reactance is reduced, which leads to increase in power transfer capacity.
- Series capacitance is inserted in the line to improve voltage profile.

TCSC, a series-controlled capacitive reactance, ensures continuous power flow. The fundamental frequency voltage is adjusted by varying the firing angle α across the fixed capacitor in a series compensated line.

Two buses i and j connected in the transmission line model with a TCSC as shown in Figure 1.15.

Two principles of the TCSC are

- Provides electromechanical damping in a large electrical system by a variable capacitive reactance.
- To avoid subsynchronous resonance, TCSC changes apparent impedance to its subsynchronous frequencies.

The equivalent model of TCSC is shown in Figure 1.16.

The objectives are achieved with TCSC using control algorithms that operate concurrently as shown in Figure 1.17.

A variable capacitor at fundamental frequency and virtual inductor at subsynchronous frequency is developed by a thyristor-controlled circuit, and it is added to the main capacitor.

The TCSC main circuit parameter λ is the quotient of the resonant frequency and the network frequency

$$\lambda = \sqrt{\frac{-X_C}{X_L}}$$

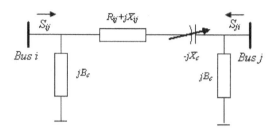

FIGURE 1.16   Model of TCSC.

The modeling of TCSC is as follows:

Let the voltages at bus i and bus j be represented by $V_i < d_i$ and $V_j < d_j$ The complex power from bus i to j is

$$S_{ij}^* = P_{ij} - Q_{ij} = V_i^* I_{ij} \tag{1.1}$$

$$= V_i^* \left[(V_i - V_j) Y_{ij} + V_i (jB_c) \right] \tag{1.2}$$

$$= V_i \left[\left[G_{ij} + j (B_{ij} + B_c) \right] - V_i^* V_j (G_{ij} + jB_{ij}) \right] \tag{1.3}$$

where,

$$G_{ij} + jB_{ij} = \frac{1}{(R_L + iX_L - iX_C)} \tag{1.4}$$

From the above equations the real and the reactive power can be written as

$$P_{ij} = V_i^2 G_{ij} - V_i V_j G_{ij} \cos(\delta_i - \delta_j) - V_i V_j B_{ij} \sin(\delta_i - \delta_j) \tag{1.5}$$

$$Q_{ij} = -V_i^2 (B_{ij} + B_c) - V_i V_j G_{ij} \sin(\delta_i - \delta_j) + V_i V_j B_{ij} \sin(\delta_i - \delta_j) \tag{1.6}$$

Correspondingly, the real and the reactive powers of bus j and i can be represented by replacing V_i by V_j.

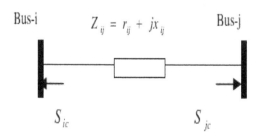

FIGURE 1.17   Injection model of TCSC.

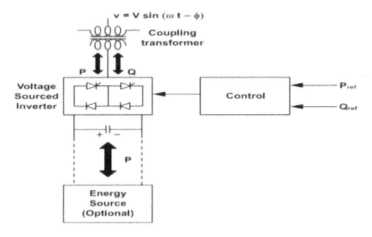

FIGURE 1.18   Static series synchronous compensator.

The real and the reactive power losses in the line are represented by Equations (1.7) and (1.8).

$$P_L = P_{ij} + P_{ji} \tag{1.7}$$

$$Q_L = Q_{ij} + Q_{ji} \tag{1.8}$$

1.11   STATIC SYNCHRONOUS SERIES COMPENSATOR (SSSC)

A SSSC, a series-controlled device that consists of a solid-state voltage source inverter coupled with a transformer that is connected in series with the line, is shown in Figure 1.18.

This device injects sinusoidal voltage in series with the line. The injected voltage is considered as an inductive or capacitive reactance connected in series with the line. This provides controlled compensation of voltage.

1.11.1   POWER FLOW IN SSSC

Active power control is needed to supply the load with rated power. Harmonics developed by reactive power flow reduce the efficiency of the line. Therefore, both real and reactive power have to be controlled.

SSSC was developed with a voltage source converter and DC storage capacitor connected in series to the line, for compensation it transfers active and reactive power to compensate voltage drops, balancing the effective X/R ratio as shown in Figure 1.19.

SSSC develops voltage opposite in phase angle to the voltage developed by the line, which guarantees fast control.

1.11.2   MODES OF OPERATION

SSSC operates in three modes as shown in Figure 1.20. A constant line reactance can be controlled with injection of voltage. Inductive reactance compensation level

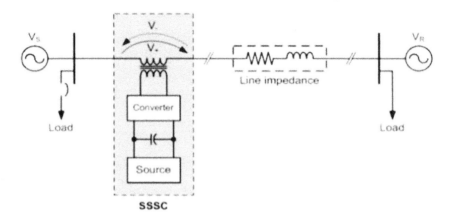

FIGURE 1.19   Power flow in SSSC.

increases from 0% to 100% for a decrease in line current. Capacitive compensation level increases from 0% to 33% for an increase in line current.

SSSC increases power in forward polarity and decreases at reverse polarity of voltage, which is fed to the line voltage drop with an increase in line impedance.

The SSSC reactance compensation effect on power flow in a transmission line can be summarized that

- for capacitive emulated reactance, active and reactive power flow increase where the effective reactance decreases in the positive direction.
- for inductive emulated reactance, active and reactive power flow decreases where the effective reactance increases in the negative direction.

1.11.3 Applications and Advantages

The SSSC corrects the voltage during fault. It has numerous benefits under normal conditions:

- Power factor correction.
- Load balancing.

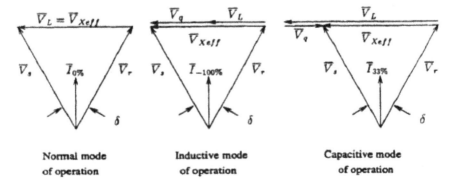

FIGURE 1.20   SSSC modes of operation.

- Power flow control.
- Reduction of harmonic distortion.

1.12 CONCLUSION

This chapter provides a detailed review of power system. The importance of power stability, including basic ideas, grouping, and meaning of related terms, was discussed. The concept of FACTS devices and their implementation for the enhancement of a power system network was well articulated in this chapter. In a more specified manner, the application of FACTS controllers in a power system network was explained.

1.13 SUMMARY

i. A detailed study about the generation, transmission, and distribution was well articulated.
ii. The power system stability, such as steady-state stability and transient stability, has been narrated in a lucid manner.
iii. Electric utility industry structure in deregulated environment was explained.
iv. FACTS devices were discussed, and the types of controllers, such as series, shunt, combined series-shunt controllers, were explained.

2 Major Issues and Constraints in Power Systems Using Traditional Methods

LEARNING OUTCOME

i. To study the importance of optimal power flow and challenges faced in electric power system.
ii. To know the subsynchronous oscillations in power systems and countermeasures.
iii. To study about the power system deregulation and congestion management.

2.1 INTRODUCTION

At present, the power industry is facing several major challenges: to offer quality and reliable electric power supply to its consumers; to minimize the generating cost, emission, and loss; and to improve voltage level. The size and operational constraints of power systems are growing as a result of huge demand for electrical energy. It is the obligation of the power system specialists to provide continuous, steady, and quality power to the buyers. These problems emphasize the need of comprehension of power system stability. Another major issue is subsynchronous resonance (SSR), a power quality issue due to expanding electrical power network. In 1970, a turbine generator at the Mohave power plant in Southern Nevada was damaged due to SSR, and again in 1971 it took time to restore the operation of it. The problem arose because of switching operations for connecting the turbine generator with a series-compensated transmission line.

There has been a vast change in techniques and procedures employed in electric power industries in the past 20 years. It has progressed from an oligopoly structure to a cutthroat environment, like transportation and telecommunications sectors. In 1980, Chile started the development of cutthroat environment for the generation of electrical power at a minimum cost price. Later in 1992, the Argentina government classified its inefficient electricity sectors into three categories: generation, transmission, and distribution. The challenge of transferring power over the existing grids has recently created an interest among the researchers and experts to develop a more robust power system by applying new technologies.

In 1968, Dommel and Tinney used Newton's power flow (PF) method to obtain minimum generating cost and losses in the power system. Later in 1972, Billinton and Sachdeva used the nonlinear programming (NLP) technique based on Powell and Fletcher algorithm. Saskatchewan Power Corporation System is considered for this study and optimized for a real power generation for a combined hydrothermal plants and reactive power optimization for individual hydro and thermal systems. In 1977, Barcelo et al. used Hessian matrix approximation based (NLP) method for real-time optimal power flow (OPF). Sparse matrix technique is used to reduce memory. This technique is applied to practical system and results are compared to Newton's PF solutions. In 1982, Housos and Irisarri used reduced Hessian matrix based on sparse technique; it is also based on Darion-Fletcher-Powell (DFF) and Broyden-Fletcher-Goldfarb-Shanno (BFGS) methods. This approach is good for a small-size power system and has convergence difficulties in large-size power systems. Shoults and Sun (1982) used real power and reactive power decomposition algorithm for OPF solutions. Real power problem is considered to minimize generating cost, and reactive power problem is used to minimize real power losses in the power system. They developed sequential unconstrained minimization technique (SUMT) to solve online OPF problem. Momoh (1999) used extended Kuhn Tucker conditions (EKT) algorithm, eigenvalues, and sensitivity methods to solve OPF. In 1989, Habibollahzadeh et al. used nonlinear method to solve OPF. To accelerate convergence, sparse matrix technique and parallel tangent methods are used. They solve OPF in two stages: in the first stage, piecewise linear approximation is used for thermal power plants; and in second stage, quadratic approximation is considered for thermal power plants to improve the convergence. In 1989, Ponrajah and Galiana used continuation method to solve OPF. Cost minimization is the objective for this quadratic cost function and linear constraints are satisfied. They used predictor corrector technique for finding minimum generating cost for various test cases. Based on the output of the results obtained in 2017 by Hongye Wang step-controlled primal dual interior point method (SCIPM), it is proved that SCIPM provides exact solutions and is able to solve large-scale market-based nonsmooth OPF. In 2008, Whei-Min Lin used the predictor-corrector interior point algorithm (PCIPA) in hybrid current power injection model to find OPF solutions. Sousa (2011) used trust region and interior-point methods based on primal dual and predictor corrector variant to solve nonlinear OPF. The results prove that the proposed algorithm has robust convergence regardless of starting point. Alejandro Pizano-Martínez (2011) proposed new transient stability constrained optimal power flow (TSC-OPF), in which the two constraints of dynamic and transient stability converted into a single stability constraint. Jacobian and Hessian matrices are used to solve the OPF problem. In 2015, Chiang et al. used interior point method to solve OPF problems subjected to security constraints. The interior point method is structure-exploiting technique, and the OPF considered is DC-OPF that solved a maximum of 500 bus system.

SSR is another major problem noticed by Anderson et al. in 1990, which occurs when series compensators are implemented in place of fixed capacitor banks in thermal generating stations. Later in 1992, Larsen provided a solution to SSR problem by using thyristor-controlled series capacitor (TCSC), thereby damping subsynchronous frequencies. In 1999, Hingorani and Gyugyi used a series-connected voltage source converter (VSC) for damping the SSR oscillations. Xueqiang (1999) presented the study of TCSC model and its prospective application in the power systems of China. The effective operation of TCSC for power swing damping issues had been demonstrated in this article. Using the Electro-Magnetic Transient Program (EMTP) digital simulation, the static and dynamic performance of TCSC had been evaluated in the power systems of China. In 2016, Seyyed Ahmad Hosseini et al. introduced a novel multiobjective solution approach for solving the transmission congestion management issues of electric power markets. Transmission congestion management plays a vital role in the deregulated electric energy markets. The voltage of power system and limits of transient stability can be considered to be avoided, thus obtaining a weak power system with low stability margins.

2.2 CONSTRAINTS IN OPTIMAL POWER FLOW

The primary objective of OPF is minimizing the transmission loss and providing stable operation and quality of power supply. Hence, OPF is a minimization problem: to minimize the generating cost of the electric power producer firm. Economic load dispatch (ELD) and unit commitment problems are also intended to minimize the generating cost related to generator and load demand, but PF problem is not considered. OPF is a combination of ELD and PF. The objective function of OPF is the same as ELD, but the equality constraint of ELD is replaced by PF condition. PF condition requires that real and reactive power generation be balanced. The emission of air pollution due to sulfur oxide, carbon oxide, and nitrogen oxide is also an important factor to be considered in power systems. Optimization problem has a solution space that is bound between many constraints, such as equality constraints and inequality constraints. Optimization technique has to find an optimal solution in the constraint-bound solution space. Power balance equation gives equality constraint for OPF problem, and is derived from load flow analysis which states that generation of real and reactive power should balance real and reactive power demand and losses. OPF problems control variables and dependent variables have lower and upper limits, and the limits on PF in the transmission lines form inequality constraints. The control variables of the problem are real power generations, generator bus voltages, and transformer tap positions; the dependent variables of the problem are reactive power generation, load bus voltages, and MVA flow in the transmission lines.

The second most important problem in power system today is that the power utilities are facing the challenge of meeting the increasing demand of electric power, and there are general remedies available such as the transmission network expansion and the creation of new generating facilities. Owing to the issues of environmental disturbance, economic considerations, and new policies, the construction of new generating plants and transmission lines is either put on hold or is avoided in many parts of the world. The flow of power in an AC transmission line is a collective function between impedance of the line, sending and receiving end voltage magnitude, and phase angle between these two end voltages. The traditional use of mechanical switched conventional series capacitors for enhancing the power transfer capability of transmission lines brought in further complications in the system by interacting with the mechanical turbine-shaft system of turbo generators. This complication results in a phenomenon known as SSR. Subsynchronous oscillation (SSO) is a dangerous event concerning instantaneous oscillations among two or more power system components such as the turbine generator, series capacitor, power electronics control, and HVDC transmission equipment control. SSO is a common expression employed to depict the field of power systems. SSO is also described as "subsynchronous interaction" (SSI), because the oscillations can be originated by numerous varieties of communications. It can be categorized into three groups: SSR, subsynchronous control interaction (SSCI), and subsynchronous torsional interaction (SSTI). One of the problems is possibility of SSR, which may lead to torsional oscillations of turbine-generator-shaft system and electrical oscillation with frequency below the subsynchronous frequency.

The other major problems faced by the electrical engineers in the deregulated power system are congestion and voltage instability. Due to increased power demand in the present modern world, the power system has to be restructured to satisfy the needs. The restructuring of power system results in maximum utilization of transmission lines in power systems. Deregulation of power system results in unbundling of generation, transmission, and distribution systems as independent companies. In such an open electric power industry, participation of private players leads to competitive power transactions. Due to such competitive power transactions between the sender and the buyer, transmission line stability and reliability eventually weakens. The participants of electricity market in some case violate the stability and thermal constraints of transmission lines by overloading the transmission lines, so-called congested lines. The occurrence of congestion will not only affect the safe and stable operation of the power network, but also have a significant impact on grid pricing and power plant bidding strategies. Hence, it is important to study and analyze congestion management.

2.3 CONVENTIONAL OPTIMIZATION METHODS IN OPTIMAL POWER FLOW

The application of traditional methods has proved helpful to a large number of researchers in the recent past. These deterministic methods are based upon mathematical programming and are used to solve different size of OPF problems. Some of the optimization methods for solving OPF problems are shown in Figure 2.1.

Apart from the five traditional methods mentioned in Figure 2.1, there is also another method called the gradient method. In the following section, a brief of all the conventional methods is presented.

2.3.1 LINEAR PROGRAMMING

A linear programming based optimization method for OPF was proposed by Mukherjee (1992). Based on the results, it could be concluded that a 220-kV five-bus system get into convergence with a single iteration itself, thereby reducing the Central Processing Unit time. Another linear programming method was presented by Chung

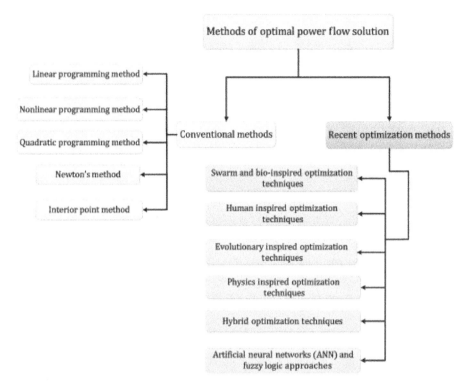

FIGURE 2.1   Optimization methods for solving OPF problems.

and Shaoyun (1997). In this method, the main motto was to reduce power loss and also find the best location for capacitor in a distribution system. Matrix inversion is not needed, thus saving computational time. Lima et al. (2003) proposed a mixed integer linear programming to find the optimal placement of thyristor-controlled phase shifter transformers in large-scale power systems. This method also claims that the computational time is less. Rau (2003) proposed a linearization method for DC-OPF with simple assumptions to yield the optimal solution. Due to its simplicity, speed, and robust nature, DC-OPF is widely used in industry.

2.3.2   GRADIENT METHOD

This method has been in use since late 1960s as a solution to OPF problems. It uses the first-order derivative vector of the objective function of a nonlinear optimization to obtain solution. Peschon et al. (1972) proposed a generalized reduced gradient (GRC) method. Penalty terms could be avoided using this method; besides, sensitivity could also be measured using a straightforward approach of computation. To summarize the benefits of using the gradient method, it is very easy to implement and it is reliable. However, it also has a drawback. Its convergence is slower compared with other methods.

2.3.3   QUADRATIC PROGRAMMING

It consists of quadratic objective function and linear constraints. Momoh proposed a model that can solve multiple objective functions. It occupies less memory and less execution time. This algorithm uses sensitivity of objective function. Due to the optimal adjustments of constraints, the method is reliable for obtaining a global optimal solution. Grudinin (1998) proposed a model for optimization of reactive power that is purely based on successive quadratic programming (SQP) method. IEEE 30 bus is considered for this method. The outcome of this approach provides quick and quality optimization compared to traditional algorithms.

2.3.4   NEWTON-RAPHSON METHOD

It is a method that considers second-order Taylor series expansion for unconstrained optimization. Real-life dispatch issues could be addressed and solved by this methodology. Chen and Chen (1997) proposed this method for solving emission dispatch in real time with sensitivity factors. Lesser execution time is an advantage. Newton Raphson method involves the Jacobian matrix. All the coefficients of the matrix can be obtained from the generalized shift distribution factor. Important indices like incremental losses and penalty factor can be obtained easily with this method. Lo et al. in his paper suggested two methods similar to the NR. Their work is based on fixed computation and change of the right vector method. The main objective is to solve issues related to line outages. These methods offer better convergence compared to Newton method and fast decoupled algorithm.

2.3.5   NONLINEAR PROGRAMMING

In this method, nonlinear objective functions are used. The constraints also remain nonlinear. Momoh et al. proposed a unique nonlinear convex network flow programming model to obtain a solution for multi-area security-constrained economic dispatch (MAED) problem. In order to solve this problem, a combination of quadratic programming and network flow programming was used. Pudjianto et al. (2002) offered LP and NLP models for allocating reactive power among generators in a deregulated environment. A faster computation speed and accuracy were achieved, yet the convergence remained a challenge.

2.3.6   INTERIOR POINT METHOD

This method is based on primal dual logarithms. It is a type of projective scaling algorithm to solve both linear and nonlinear optimization problems. Granville (1994) proposed the interior point method for obtaining a solution to reactive power dispatch. G.L. Torres proposed a solution to OPF problem using rectangular form. Castronuovo et al. (2001) changed Interior Point Method into an upgraded version for solving OPF. Further, advances were made in the same as proposed by Hua Wei et al. He used perturbed KKT condition to solve the OPF. It is not sensitive to the size of network. Table 2.1 lists a summary of some literature works based on optimization and Figure 2.2 shows a pictorial representation of existing hybrid algorithms.

TABLE 2.1

Summary of Some Literature Works Related to Optimization

Objective Function Used	Method of Optimization	Control Variable for Optimization	Type of the System
Minimize real power loss	Particle swarm optimization	Q of PV, P and Q of BESS, CL, tap positions of transformer	Distribution
Minimize real power loss	Ant colony optimization	Generator bus voltages, tap positions of transformer, Q of capacitor banks	Transmission
Minimize real power loss	ES	Generator bus voltages, tap positions of transformer, Q of capacitor banks	Transmission
Minimize total cost of a distribution system	PSO	Q of PV, Q of EV	Distribution
Minimize real power loss	SO (SOCP)	Q of PV, subject to stochastic P of PV	Distribution

2.4 SUBSYNCHRONOUS RESONANCE ANALYSIS METHOD AND COUNTERMEASURES

For analyzing the SSR effect, the following analytical approaches are available:

- Routh-Hurwitz criterion
- Time domain simulation analysis
- Frequency scanning
- Eigenvalue analysis

2.4.1 ROUTH-HURWITZ CRITERION

Routh's stability criteria approach provides information only on absolute stability of the system and not on the relative stability of the system. Also, this approach does not provide suitable method for improving the stability. It does not offer flexibility with respect to subsystems like excitation systems, speed governing systems, etc. This Routh-Hurwitz criterion cannot describe the complete behavior of SSR oscillations and also various modes of oscillations present in the system. Hence, this approach is not suitable to analyze the SSO present in the series-compensated systems.

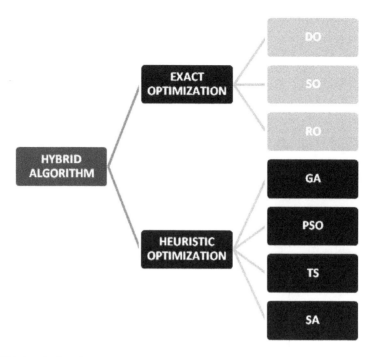

FIGURE 2.2   A pictorial representation of existing hybrid algorithms.

In Figure 2.2, the following abbreviations were used:

DO: deterministic optimization
SO: stochastic optimization
RO: robust optimization
GA: genetic algorithm
PSO: particle swarm optimization
TS: Tabu search
SA: simulated annealing

2.4.2   Frequency Scanning

The frequency scanning method is another simple technique for analysis of SSR that utilizes phase and gain margin indicators for finding the absolute and relative stabilities: by calculating the driving point impedance at the desired frequency range as viewed from the neutral bus of generator. The main drawback of this method is that it can be used only for preliminary SSR studies in large networks.

2.4.3   Eigenvalue Analysis

The eigenvalue analysis approach is more accurate in comparison with frequency scanning for identifying SSR. The procedure for eigenvalue analysis includes

1. Modeling of power system network
2. Modeling of generator electrical circuits

TABLE 2.2

Association Between Eigenvalues and the Characteristics

S. No	Type of Eigenvalue	Implication of Mode
1	Real value	Nonoscillatory mode
2	Negative real	Decaying mode
3	Large absolute negative real	Fast decaying mode
4	Positive real	Oscillatory with increasing magnitude

3. Modeling of turbine generator spring mass system
4. Eigenvalue calculation of the interconnected systems

The oscillatory modes are predicted from the eigenvalues. Table 2.2 provides a clear-cut understanding of the eigenvalues and their interpretation.

In general, a positive real component implies that the characteristics will be oscillatory in nature with increasing amplitude, whereas a negative real component indicates damped type of oscillatory behavior.

2.4.4 TIME DOMAIN SIMULATIONS

By using this method, the risk can be assessed for all types of faulty conditions. This method is most suitable for studies related to torques during transients. The time limit for simulations must be higher for analyzing the absorbance of SSR in steady-state condition. The confirmation of results obtained from the already obtained eigenvalues can be successfully done by using this approach.

2.4.5 COUNTERMEASURES OF SSR

The basic measures for countering SSR can be grouped into three main categories, as depicted in Figure 2.3.

The various countermeasures of SSR are as follows.

2.4.5.1 Pole-Face Amortisseur Windings

The net negative resistance of the generator can be minimized at frequencies lesser than those occurring at synchronous speed by addition of these windings. Hence, net positive resistance will be produced at the frequencies at slip condition. This method stands as an inexpensive one for installation in new machines, but it can't be practically implemented in existing machines.

2.4.5.2 Static Blocking Filters

This high Q filter is used to block the electrical resonance current of the transmission system from entering the generator, which may interact with the torsional modes, tuned to the torsional mode natural frequencies, and inserted between the star-connected high-voltage winding of the setup transformer and the ground. When electrical resonance corresponding to any one torsional mode natural frequency begins

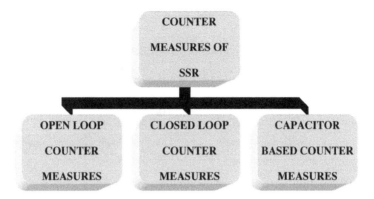

FIGURE 2.3   Categories of countermeasures of SSR.

to develop, the impedance of one of the mode filters will become extremely large to prevent further growth of the electrical resonance current, but it will be small at other frequencies. The main problem in the blocking filter design is to maintain constant filter parameter values in an environment of drastic changes in temperature between day and night. The cost of the filter units are high due to high insulation level of transformer as it is used for high-voltage and large-current application.

2.4.5.3   Dual Gap Flashing

This dual gap flashing scheme of capacitors has been developed to protect the generating unit itself from shaft damage due to SSR in a system. Air gaps parallel to the capacitor will flash over at lower current level of 2.2 pu to reduce a transient torque impact to the generator shaft and the current level will be reset each time after flashing to about 3–3.6 pu to allow a current decay to the level for successful reinsertion of the series capacitor.

2.4.5.4   Supplementary Excitation Control

The excitation control of SSR is less expensive in comparison with static fillers because it is the low energy side that is being controlled. Power system stabilizer type of control is found to be more difficult due to several controls deigned for individual torsional modes of different frequencies, which will interfere with each other, known as modal interaction.

2.4.5.5   Reactive Power Control

It is a very important aspect in damping SSR oscillations. When the generator terminal is connected by a static var compensator (SVC) by generator, speed deviation or modal speed deviation as the controlling signal is proven to damp the SSR oscillations.

2.4.5.6   Thyristor-Controlled Series Capacitor

TCSC replaces traditional mechanically controlled series compensators. There is an increase in dynamic and transient stabilities. These higher values aid in control of load flows flexibly, thus mitigating the oscillations by SSR.

2.5 ELECTRICITY MARKET DEREGULATION AND CONGESTION

In the current scenario energy market, the deregulation problems and power quality issues are the major issues in the present power systems throughout the world. With deregulation of the power sector, load forecasting is essential not only for system operators but also for market operators, private energy sectors, and other energy market participants, so that sufficient energy dealings can be scheduled and suitable operational plans and bidding strategies could be organized.

2.5.1 OBJECTIVES OF THE DEREGULATED POWER SECTORS

- To deliver electricity for all reasonable demands.
- To boost the competition in the generation and supply of electricity.
- To improve the continuity of supply and the quality of services.
- To promote efficiency and economy of the power system.

2.5.2 USES

- No disturbances in generation as activities of transmission and distribution are separated.
- Number of generation industries will be increased, thereby transmitting power using the same network
- Better quality and reliability of power for the consumers.
- Fifty percent reduction in electricity prices and reduced service losses.
- Removal of monopoly of local electric utility.
- Increased competition among the power producers.

2.5.3 TRADITIONAL METHODS OF INCREASING POWER TRANSFER CAPABILITY

- Replacing the line conductor with larger size for carrying high power and replacing terminal equipment.
- Improvement of operating voltage of a transmission line.
- To eliminate overload in transmission lines by providing additional path in transmission lines.

2.5.4 CONGESTION MANAGEMENT

Congestion management is another important tasks performed by power system operators to confirm the operation of transmission system within operating limits. In the present electric power market, it becomes extremely important and it is an obstacle to the electricity trading.

The congestion management methods can be categorized as follows:

- Sensitivity factors based methods
- Auction-based congestion management
- Pricing-based methods
- Re-dispatch and willingness to pay methods

The congestion management increases in terms of price in the current environment of deregulated system, and it becomes barrier in the electricity trade mart. The FACTS devices are widely used to reduce the congestion and smoothen locational marginal price (LMP). The popular FACTS devices involving in the congestion management problem in the power system are TCSC, TCPAR, and UPFC, and these are used for redirecting the power beginning from congested path to other path.

2.5.4.1 Power Flow and Balancing Control

In controlling the PF in the transmission and distribution lines, the FACTS controllers play a vital role. In view of this, TCSC, SSSC, and UPFC are used to enable the PF in the transmission lines and it made possible in efficient utilization power at the end user.

2.5.4.2 Dynamic Applications

With respect to dynamic applications, FACTs controllers are making significant contribution toward enhancing stability in the transient condition by giving fast and rapid response. During emergency conditions, the oscillation damping dynamic voltage control is used to improve the system from voltage collapse and mitigation of SSR. These devices can also be used for interconnecting power network for expanding the power network over a region with long distance.

2.5.4.3 Loading Margin Improvement

Currently, the fundamental reason for worldwide block outs is the voltage collapse occurring in the maximum load ability. Therefore, the shunt compensators are used to transfer the power at the maximum capability in the power system efficiently.

2.6 OPTIMAL POWER FLOW PROBLEM FORMULATION

The first step is to form the objective function. It can include any parameter optimization like minimizing fuel cost or active power losses or any desired criteria. Based on the generated objective function, the formulation varies. For example, if the fuel cost is to be minimized, then the following procedure can help to deduce the problem formulation. The quadratic cost function is the most commonly used one, where the input to the units can be either in terms of total heat energy per hour or total fuel cost per hour. So after writing the technical details in terms of equations, an approach to solve these is chosen. The solution can be obtained from any of the methods discussed earlier.

2.7 CONCLUSION

In this chapter, major power system problems were discussed and a basic understanding of OPF was achieved. The major constraints involved in transmitting PF are discussed. The various optimization methods used for solving OPF problem were discussed in detail. The problem of SSR was brought to the limelight and the various traditional measures to overcome them were also discussed in detail. The importance of deregulated power system and congestion management was discussed. Finally, an outline of formulation of the OPF problem was presented to make the readers understand the various parameters to be considered in forming the objective function. The objective function is the root of the problem under consideration and only with the help of it, a proper approach for solving it can be identified. Hence, this chapter serves to aid in developing an OPF problem formulation and solving it.

2.8 SUMMARY

    i. The major constraints involved in OPF were discussed.

    ii. Summary of traditional approaches applied for solving OPF with literature survey was presented.

    iii. Analysis of SSR and countermeasures was presented.

    iv. The formulation of OPF was presented.

REFERENCES

Ambriz-Pérez, H, Acha, E, Fuerte-Esquivel, CR & De la Torre, A, 1998, 'ncorporation of a UPFC model in an optimal power flow using Newton's method', IEE Proceedings: Generation, Transmission and Distribution, vol. 145, no. 3, pp. 336–344.

Anderson, PM, Agrawal, BL & Van Ness, JE, 1990, Subsynchronous Resonance in Power Systems, IEEE Press.

Barcelo, WR, Lemmon, WW & Koen, HR, 1977, 'Optimization of the real time dispatch with constraints for secure operation of bulk power systems', IEEE Transactions on Power Apparatus and Systems, vol. 96, no. 3, p. 7.

Billinton, R & Sachdeva, SS, 1972, 'Optimal real and reactive power operation in a hydro thermal system', IEEE Transactions on Power Apparatus and Systems, vol. 91, pp. 1405–1411.

Castronuovo, ED, Campagnolo, JM & Salgado, R, 2001, 'New versions of interior point methods applied to the optimal power flow problem', IEEE Transactions on Power Systems, vol. 16., pp. 1–6.

Chen, SD & Chen, JF, 1997, 'A new algorithm based on the Newton-Raphson approach for real-time emission dispatch', Electric Power Systems Research, vol. 40, pp. 137–141.

Chiang, N & Grothey, A, 2015, 'Solving security constrained optimal power flow problems by a structure exploiting interior point method', Optimization and Engineering, vol. 16, no. 1, pp. 49–71.

Chung TS & Shaoyun Ge, 1997, 'A recursive LP-based approach for optimal capacitor allocation with cost-benefit consideration', Electric Power System Research, vol. 39, pp. 129–136.

Dommel, HW & Tinney, WF, 1968, 'Optimal power flow solutions', IEEE Transactions on Power Apparatus and Systems, vol. 87, no. 10, pp. 1866–1876.

Granville, S, 1994, 'Optimal reactive dispatch through interior point methods', IEEE Transactions on Power Systems, vol. 9, no. 1, pp. 136–146.

Grudinin, N, 1998, 'Reactive power optimization using successive quadratic programming method', Transactions on Power Systems, vol. 13, no. 4, pp. 1219–1225.

Habibollahzadeh, H, 1989, 'Hydrothermal optimal power flow based on a combined linear and non-linear programming methodology', IEEE Transactions on Power Systems, vol. 4, no. 2, pp. 530–537.

Hingorani, NG & Gyugyi, L, 1999, Understanding FACTS: Concepts and Technology of Flexible AC Transmission System, IEEE Press, New York.

Housos, EC & Irisarri, GD, 1982, 'A sparse variable metric optimization applied to the solution of power system problems', IEEE Transactions on Power Apparatus and Systems, vol. 101, pp. 195–202.

Larsen, EV, Bowler, CEJ, Damsky, B & Nilsson, S, 1992, 'Benefits of Thyristor Controlled Series Compensation', in CIGRE Annual Meeting, Paris, Paper No. 14/37/38-04.

Lima, FGM, Galiana, FD, Kockar, I & Munoz, J, 2003, 'Phase shifter placement in large-scale systems via mixed integer linear programming', IEEE Transactions on Power Systems, vol.18, no. 3, pp. 1029–1034.

Lo, KL & Meng, ZJ, 2004, 'Newton-like method for line outage simulation', IEE Proceedings – Generation, Transmission and Distribution, vol. 151, no. 2, pp. 225–231.

Molzahn, DK, 2013, 'Implementation of a large-scale optimal power flow solver based on semi definite programming', IEEE Transactions on Power Systems, vol. 28, no. 4, pp. 3987–3998.

Momoh, JA, 1989, 'A generalized quadratic-based model for optimal power flow', CH2809-2/89/0000-0261 $1.00, © IEEE, pp. 261–267.

Momoh, JA, 1999, 'A review of selected optimal power flow literature to 1993. Part I: Nonlinear and quadratic programming approaches', IEEE Transactions on Power Systems, vol. 14, no. 1, pp. 96–104.

Momoh, JA, & EI-Hawary, ME, 1999, 'A review of selected optimal power flow literature to 1993 Newton, linear programming and interior point methods', IEEE Transactions on Power Systems, vol. 14, no. 1, pp. 105–111.

Mukherjee, SK, 1992, 'Optimal power flow by linear programming based optimization', Southeastcon '92, Proceedings, IEEE, vol. 2, 527–529.

Peschon, J, Bree, D & Hajdu, L, 1972, 'Optimal power-flow solutions for power system planning', Proceedings of IEEE, vol. 6, no. 1, 64–70.

Pizano-Martínez, A, 2011, 'A new practical approach to transient stability-constrained optimal power flow', IEEE Transactions on Power Systems, vol. 26, no. 3, pp. 1686–1696.

Pudjianto, D, Ahmed, S & Strbac, G, 2002, 'Allocation of VAR support using LP and NLP based optimal power flows', IEE Proceedings: Generation, Transmission and Distribution, vol. 149, no. 4, pp. 377–383.

Rau, N, 2003, 'Issues in the path toward an RTO and standard markets', IEEE Transactions on Power Systems, vol. 18, no. 2, 435–443.

Shoults, RR & Sun, DT, 1982, 'Optimal power flow based on P-Q decomposition', IEEE Transactions on Power Apparatus and Systems, vol. 101, pp. 397–405.

Sousa, AA, 2011, 'Robust optimal power flow solution using trust region and interior-point methods', IEEE Transactions on Power Systems, vol. 26, no. 2, pp. 487–499.

Wang, H, 2007, 'On computational issues of market-based optimal power flow', IEEE Transactions on Power Systems, vol. 22, no. 3, pp. 1185–1193.

Wei, H, Sasaki, H, Kubokawa, J, and Yokoyama, R. Large scalehydrothermal optimal power flow problems based on interior point non-linear programming. IEEE transactions on power systems, vol. 15, no. 1, pp. 396–403.

Xueqiang, 1999, 'Study of TCSC model and prospective application in the power systems of China', Proceedings of the IEEE 1999 International Conference on Power Electronics and Drive Systems. PEDS'99 (Cat. No.99TH8475).

3 Particle Swarm Optimization Based Optimal Power Flow

LEARNING OUTCOME

i. To study the importance of particle swarm optimization (PSO) algorithm in optimal power flow.
ii. To compare the total power generation and demand using PSO algorithms versus various algorithms.
iii. To make a comparative study on fuel cost with various optimization algorithms.
iv. Comparative analysis of real power losses by different algorithms.

3.1 INTRODUCTION

The rising trend in demand of electricity forces power networks to function close to their maximum bounds, thereby resulting in increasing transmission losses and contaminants produced by the power sector. Hence, the power system operators must find the most economical and secure combination of generation units, which can be accomplished by using a powerful optimal power flow (OPF) tool. OPF is an optimizing tool for power system operation analysis, scheduling, and energy management. OPF tool is being widely used because of its capabilities to deal with various situations. This problem involves the optimization of an objective function that can take various forms while satisfying a set of operational and physical constraints. This chapter presents about the particle swarm optimization (PSO) technique for solving OPF problem. Recent research works indicate that intelligent algorithms have been used to solve OPF. The various intelligent algorithms are genetic algorithm (GA), simulated annealing (SA), ant colony algorithm (ACA), bee algorithm (BA), differential evolution (DE), PSO, harmony search (HS), firefly algorithm (FFA), cuckoo search algorithm (CS), and flower pollination algorithm (FPA). In 2002, Bakirtzis et al. adapted enhanced genetic algorithm (EGA) to solve OPF. Later in 2010, Yan and Li used improved DE algorithm to solve OPF; Malik and Srinivasan (2010) used GA to solve OPF for which they considered elitism and nonuniform mutation rate. Attia et al. (2012) used adapted GA in which the population size is altered based on fitness function for solving OPF. They assert fast convergence better solution to OPF. Prathiba et al. (2014) used FPA to solve economic emission dispatch problem. Azizipanah-Abarghooee et al. (2014) used a modified shuffled frog leaping algorithm (MSFLA) to solve OPF with the flexible alternating current transmission system (FACTS) devices. Jordehi (2015) used brainstorm optimization algorithm (BSOA) to find the optimal size and location of the FACTS devices. In their work, they considered static var compensator (SVC) and thyristor-controlled series capacitor (TCSC) to reduce the losses in the system and to improve the voltage in the power system.

The objective of the OPF problem is to minimize the generating cost. The objective cost function is the function of real power generation of the committed generator without valve point loading effect, as given in Equation (3.1). The modulus term with coefficients d_i and e_i are added for considering valve point effect, as given in Equation (3.2).

Minimize $F_1(x)$ as follows:

$$\text{Minimize } F_1(x) = \sum_{I=1}^{NG} \left(a_i P_{gi}^2 + b_i P_{gi} + c_i \right) \$/\text{hour} \tag{3.1}$$

(or)

$$\text{Minimize } F_1(x) = \sum_{I=1}^{NG} \left(a_i P_{gi}^2 + b_i P_{gi} + c_i \right) + \mid d_i \, \sin\left(e_i \left(P_{gi\_min} - P_{gi} \right) \right) \mid \$/\text{hour} \tag{3.2}$$

Equations (3.3)–(3.8) give the control variables of the objective function:

$$x = \left[P_g, V_g, T, Q_c \right] \tag{3.3}$$

$$P_g = \left[P_{g1}, P_{g2}, \ldots, P_{NG-1} \right] \tag{3.4}$$

$$V_g = \left[V_{g1}, V_{g2}, \ldots, V_{NG} \right] \tag{3.5}$$

$$T = \left[T_1, T_2, \ldots, T_{NT} \right] \tag{3.6}$$

$$Q = \left[Q_{C1}, Q_{C2}, \ldots, Q_{CNC} \right] \tag{3.7}$$

$$NCV = ((NG-1) + NG + NT + N_{UPFC} \tag{3.8}$$

where
  $F_1(x)$ is the fuel cost of generation
  x is the list of control variables
  a_i, b_i, c_i, d_i, and e_i are the coefficients of fuel cost
  P_g is the real power generation
  V_g is the voltage magnitude of generator bus
  T is the transformer tap position
  Q_c is the reactive power support in the bus

NG is the number of generator

NT is the number of transformer

N_{UPFC} is the number of UPFC

NCV is the number of control variables

In the second objective function, the environmental problems of gaseous pollution by the thermal power plants are also considered. Hence, the emission minimization $F_2(x)$ objective function is given by Equation (3.9):

$$F_2(x) = \sum_{I=1}^{NG} \left(\gamma_i P_{gi}^2 + \beta_i P_{gi} + \alpha_i + \xi_i e(\lambda_i P_{gi}) \right) \$ / \text{hour} \tag{3.9}$$

where

Υ, β, α, ξ, and λ are emission coefficients.

The third objective function is loss minimization. A conductor is used to transmit power from the generating station to the consumer. However, the conductor has some resistance and it consumes power as heat losses. Hence, the reduction in loss reduces the generating cost. This loss minimization $F_3(x)$ forms the objective function, and it is given in Equation (3.10).

$$F_3(x) = \sum_{I=1}^{nbr} g_k \left[V_i^2 - V_j^2 - 2V_i V_j \cos\left(\theta_i - \theta_j\right) \right] \text{MW} \tag{3.10}$$

where

nbr is the number of branch or transmission line

g_k is the conductance of the conductor

V_i is the sending end bus voltage magnitude

V_j is the receiving end bus voltage magnitude

θ_i, θ_j are the sending and receiving end voltage angles

Voltage stability is another important factor to be considered for reliable operation of power system. This voltage stability can be measured using L-index. To attain maximum stability, the L-index should be minimum. This L-index minimization is considered as fourth objective function, and it is given in Equation (3.11).

L-index minimization $F_4(x)$ is given as

$$L_j = |1 - \sum_{i=1}^{ng} F_{ji} \frac{V_i}{V_j}| \tag{3.11}$$

The matrix F_{ji} is given in Equation (3.11).

$$[F] = -[Y_{LL}]^{-1}[Y_{LG}] \tag{3.12}$$

In Equation (3.12), Y_{LL} is the submatrix of Y_{bus} for all load buses in the system. The matrix Y_{LG} is the submatrix of Y_{bus}, which is the corresponding generator bus linked to the load buses. The current equation for this admittance matrix is given in Equation (3.13):

$$I_{bus} = Y_{bus}V_{bus} \tag{3.13}$$

The current equation can be written in submatrix form as given in Equation (3.14)–(3.16):

$$\begin{bmatrix} I_L \\ I_G \end{bmatrix} = \begin{bmatrix} Y_{LL} & Y_{LG} \\ Y_{GL} & Y_{GG} \end{bmatrix} \begin{bmatrix} V_L \\ V_G \end{bmatrix} \tag{3.14}$$

$$I_L = Y_{LL}V_L + Y_{LG}V_G \tag{3.15}$$

$$Y'_{LL} \cdot V_L + Y_{LG} \cdot V_G = 0 \tag{3.16}$$

With reference to the above equations, it is found that the load bus voltage depends on generator bus voltage and on admittance of the line connecting the generator bus to the load bus. The dependency of load bus voltage is given in Equations (3.17) and (3.18):

$$V_L^k = \left(\left(Y'_{LL}{}^{-1} \right) \cdot Y_{LG} \right)_{k,i} \cdot V_g^k \tag{3.17}$$

$$L_j = \sum_{i=1}^{NG} \left(Y'_{LL} \cdot Y_{LG} \right)_{j,i} \tag{3.18}$$

The fourth objective function F_4 derived from the L-index is given in Equation (3.19):

$$F_{4(x)} = L = \max\left(L_j \right) \tag{3.19}$$

The above multiobjective OPF problem is subjected to constraints on control and dependent variables. These constraints are grouped into equality and inequality constraints. For this problem, power balance equation gives equality constraint, as given in Equations (3.10) and (3.21). Equation (3.20) is the equality constraint for real power, and Equation (3.21) is the equality constraints for reactive power.

Power balance equation gives equality constraint for OPF problem. This power balance equation is derived from load flow analysis.

$$\sum_{i=1}^{NG} P_{Gi} = P_D + P_L \tag{3.20}$$

$$\sum_{i=1}^{NG} Q_{Gi} = Q_D + Q_L \tag{3.21}$$

where
P_G is the real power generation, MW
Q_G is the reactive power generation, MVAr
P_D is the real power demand, MW
Q_D is the reactive power demand, MVAr
P_L is the real power loss, MW
Q_L is the reactive power loss, MVAr

The limits on control and dependent variables are derived from inequality constraint. Control variable, P_g, has its minimum and maximum limits for power generation, as given in Equation (3.22). Control variable, reactive power generation Q_g has its minimum and maximum form inequality constraints given in Equation (3.23). Similarly, minimum and maximum limits on bus voltage magnitude, transformer tap positions, MVA limits of transmission line, and UPFC form inequality constraints given in Equations (3.24–3.27).

3.1.1 CONTROL VARIABLES

Control variables include real power generations, generator bus voltages, and transformer tap positions.

3.1.2 DEPENDENT VARIABLES

Dependent variables include reactive power generation, load bus voltages, and *MVA* flow in the transmission lines, which are stated as follows:

$$P_{gi}^{\min} \le P_{gi} \le P_{gi}^{\max} \text{ for } i = 1 \text{ to } NG \tag{3.22}$$

$$Q_{gi}^{\min} \le Q_{gi} \le Q_{gi}^{\max} \text{ for } i = 1 \text{ to } NG \tag{3.23}$$

$$V_i^{\min} \le V_i \le V_i^{\max} \text{ for } i = 1 \text{ to } NB \tag{3.24}$$

$$T_i^{\min} \le T_i \le T_i^{\max} \text{ for } i = 1 \text{ to } NT \tag{3.25}$$

$$MVA_i \le MVA_i^{\max} \text{ for } i = 1 \text{ to } Nbr \tag{3.26}$$

$$UPFC_i^{\min} \le UPFC_i \le UPFC_i^{\max} \text{ for } i = 1 \text{ to } N_{UPFC} \tag{3.27}$$

where
$P_{gi}^{\min}$ is the lower limit of real power generation
$P_{gi}^{\max}$ is the upper limit of real power generation

$Q_{gi}{}^{min}$ is the lower limit of reactive power generation
$Q_{gi}{}^{max}$ is the upper limit of reactive power generation
P_{gi} is the real power generation
Q_{gi} is the reactive power generation
V_{imin} is the minimum bus voltage limit
V_{imax} is the maximum bus voltage limit
V_i is the bus voltage

3.2 PSO ALGORITHM

PSO is a stochastic optimization technique developed in 1995 by Kennedy and Eberhart (1995), inspired by social behavior of bird flocking. In birds flocking, the bird that has high energy leads the flock and the bird that has least energy follows the flock at the end. This gives an idea to find global best – highest fitness and local best that is the best of every individual. The position of each bird is updated by adding a new velocity with its old position. The main strength of PSO is its fast convergence, compared with many global optimization algorithms like GA, SA, and other global optimization algorithms.

PSO has many similarities with evolutionary computation techniques such as GA. The system is initialized with a population of random solutions and searches for optima by updating generations. PSO has no evolution operators such as crossover and mutation. In PSO, the potential solutions called particles fly through the problem space by following the current optimum particles. PSO is a global optimization algorithm for dealing with problems in which a best solution can be represented as a point or surface search in n-dimensional space.

In past several years, PSO has been successfully applied in many research and application areas. It is demonstrated that PSO gets better results in a faster and cheaper way compared with other methods.

Another reason that PSO is attractive is that there are few parameters to adjust. One version, with slight variations, works well in a wide variety of applications. PSO has been used for approaches that can be used across a wide range of applications, as well as for specific applications focused on a specific requirement.

PSO algorithm evaluates each particle in the population to find its fitness and to find gbest (global best) and pbest (personal best).

The general form of PSO is given in Equations (3.28)–(3.30):

$$\text{Minimize } f = \lambda_1 f_1 + \lambda_2 f_2 \tag{3.28}$$

$$\text{Subject to: } g(|V|, \delta) = 0 \tag{3.29}$$

$$X_{min} \leq X \leq X_{max} \tag{3.30}$$

where
  f_1 is the total generating cost without valve point effect in \$/hour, as given in Equation (3.1)
  f_2 is the loss in MW as given in Equation (3.10)

λ_1, λ_2 are the penalty factors corresponding to each objective

$g(|V|, \delta)$ is the power flow balance equation

X is a set of control variables

X_{min}, X_{max} are the minimum and maximum values of control variables

Control variables in the problem are identified as X, and then population size, constants c_1, c_2, and weight w are decided before encoding the PSO algorithm. The flowchart of PSO is shown in Figure 3.1.

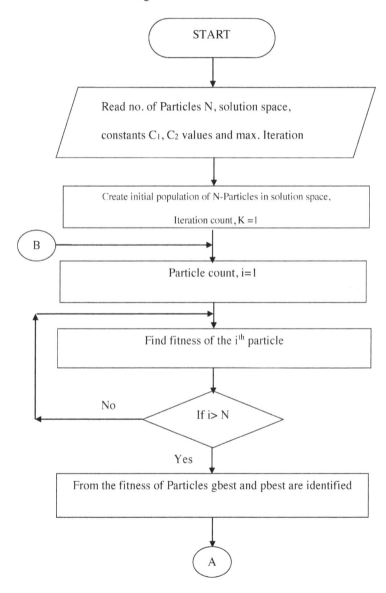

FIGURE 3.1   PSO algorithm flowchart.

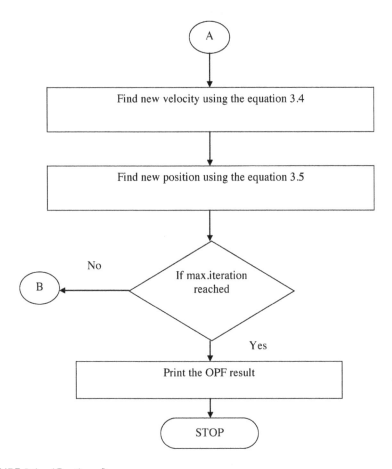

FIGURE 3.1   (*Continued*)

3.3   PSO-BASED OPF

For optimizing the OPF, the control variables considered are real power generation, generator bus voltages, and transformer tap position. The limits on these control variables form prime constraints in addition to power balance condition. The process of implementing PSO includes finding pbest, gbest, new velocity, and new position.

3.3.1   ENCODING

Encoding is the process of converting objective function into fitness function and decision variable into particle.

3.3.2   FITNESS FUNCTION

PSO evaluates fitness function for each particle in the population. Based on fitness, gbest and pbest of the particle are identified. Objective function value for a chromosome is called fitness for the particle.

3.3.3 NEW VELOCITY

Using the values of gbest and pbest, new velocity is calculated from old velocity. Equation (3.31) gives the new velocity.

$$v_i^{k+1} = v_i^{k+1} + c_1 * rand_1 * (pbest_i - x_i) + c_2 * rand_2 * (gbest_i - x_i) \quad (3.31)$$

where
  V_i is the velocity of the ith particle
  K is the iteration number
  c_1, c_2 are the constants
  $rand$ is the random number
  X_i is the position of the ith particle
  $pbest$ is the personal best of the particle till current iteration
  $gbest$ is the global best among all the particles

3.3.4 NEW POSITION

New position is obtained from the old position, and the new velocity is as given in Equation (3.32):

$$x_i^{k+1} = x_i + v_i^{k+1} \quad (3.32)$$

where
  x_i is the position of the ith particle
  v_i^{k+1} is the new velocity
  K is the iteration number

3.3.5 STOPPING CRITERIA

PSO improves problems' solution through iteration by iteration, and the iteration has to be stopped when either the problem is converged or the iteration has reached its minimum value. Stopping of iteration is important to provide solution for time complexity.

3.4 PSO ALGORITHM FOR SOLVING OPF

The steps involved in PSO algorithm to solve OPF are as follows:

1. Nominate control variables of OPF as gene and chromosome.
2. Create initial population.
3. Find the fitness of particles.
4. Find the gbest and pbest.
5. Find the new velocity.
6. Find the new position.
7. Repeat steps 3–6 until stopping criterion is fulfilled.
8. Print the optimal result after stopping criterion is satisfied.

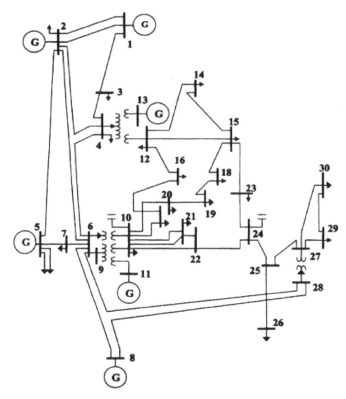

FIGURE 3.2   IEEE 30 bus system single-line diagram. (*Source:* Power Systems Test Case
Archive, University of Washington Electrical Engineering.)

3.5   NUMERICAL RESULTS AND DISCUSSION

MATLAB has been used for simulation. Dedicated software is developed in MATLAB
for this optimization problem and the results are discussed in this section. IEEE 30 bus
system is considered and shown in the Figure 3.2. For IEEE 30 bus system, five real
power generation, six generator voltage, and four transformer tap positions are used as
control variables. In this simulation, OPF without valve point loading effect and 100%
loading effects are considered. For the multiobjective of generating cost without valve
point effect as given in Equation (3.1) and for loss, Equation (3.10) is used.

3.5.1   PARAMETERS OF PSO ALGORITHM

Parameter	Value
Number of particles	60
c_1	2.05
C_2	2.05
Weight factor w	1.2
No. of iterations	200

TABLE 3.1
PSO-Based Multiobjective OPF Solution

S. No.	Variables	P_{min} (MW)	P_{max} (MW)	MSFLA (2011)	ABC Algorithm (2015)	BBO Algorithm (2016)	PSO-OPF
1	P_{G1} (MW)	50	200	179.1929	177.236	176.611	176.12
2	P_{G2} (MW)	20	80	48.9804	48.701	48.624	47.03
3	P_{G5} (MW)	15	50	20.4517	21.353	21523	22.15
4	P_{G8} (MW)	10	35	20.9264	21.134	21.825	20.64
5	P_{G11} (MW)	10	30	11.5897	11.898	12.170	12.27
6	P_{G13} (MW)	12	40	11.9579	12.00	12.271	13.72
7	Total power generation, MW	—	—	293.0991	292.322	293.024	291.930
8	Total demand, MW	—	—	283.4	283.4	283.4	283.4
9	Real power loss, MW	—	—	9.6991	8.934	9.624	8.530
10	Generating cost ($/hour)	—	—	802.287	800.0963	802.721	799.454

Table 3.1 enlists the total power generation, total demand, real power lost, and generation cost. It is observed from the table that the total power generated by the MSFLA is 293.0991, whereas with artificial bee colony (ABC) and biogeography-based optimization (BBO) algorithms the values are 292.322 and 293.024, respectively. With the proposed PSO algorithm, the total power generated was 291.930, which is comparatively lower than all three algorithms. The next component is real power loss with MSFLA, which is 9.6991 MW, and with ABC and BBO algorithms, the power loss is 8.934 and 9.624 MW, respectively. By applying PSO algorithm, there is considerable reduction in power loss of 8.530 MW.

Figure 3.3 shows the loss minimization convergence curve, which is the curve between losses and iterations, and the cost is converged at 27th iteration. Figure 3.4 shows the fuel cost versus iteration curve for IEEE 30 bus system and the cost is converged at 43rd iteration, which shows the effectiveness of the algorithm. The time taken to reach the convergence is 24 seconds.

Figure 3.5 shows the total generation and demand for the base case and proposed PSO method. The total power generation is reduced to 291.930 MW, which in turn reduces the power losses in the system. Figure 3.6 shows the comparative real power loss between the existing and proposed PSO algorithms. It has been observed that loss is reduced to 8.53 MW. Figure 3.7 shows fuel cost comparison between the existing and proposed PSO algorithms and it is also observed that fuel cost is reduced to $799.454/hour.

The PSO algorithm is executed for 50 trails and the corresponding worst, average, and best results are shown in Figure 3.8. Figure 3.8 shows the fuel cost optimization, and Figure 3.9 shows the real power loss optimization. By using PSO algorithm, the fuel cost is reduced by 1% and real power loss is reduced by 12.05%.

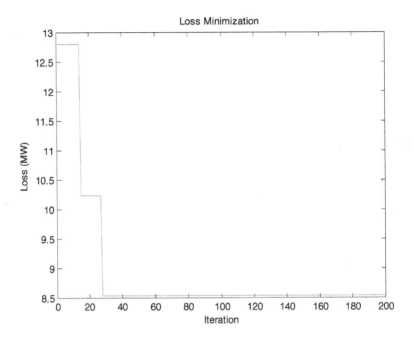

FIGURE 3.3   Losses versus iteration curve for IEEE 30 bus system.

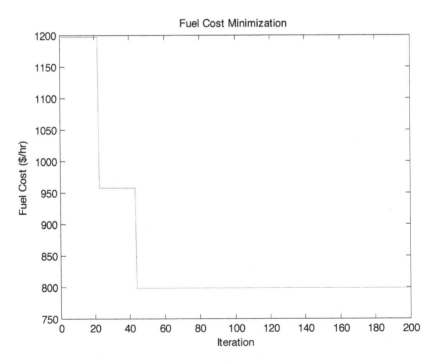

FIGURE 3.4   Fuel cost versus iteration curve for IEEE 30 bus system.

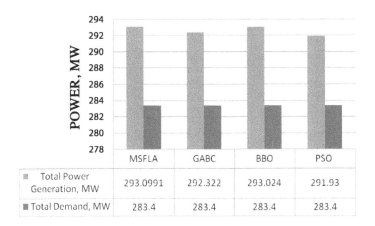

FIGURE 3.5 Comparison of total power generation and demand versus various algorithms.

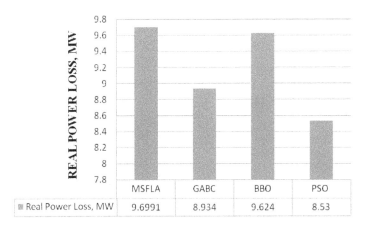

FIGURE 3.6 Comparison of real power losses versus various algorithms.

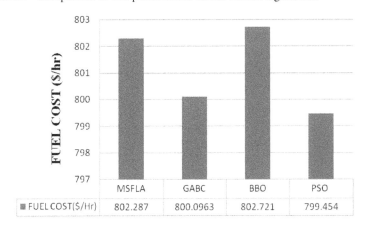

FIGURE 3.7 Comparison of fuel cost versus various algorithms.

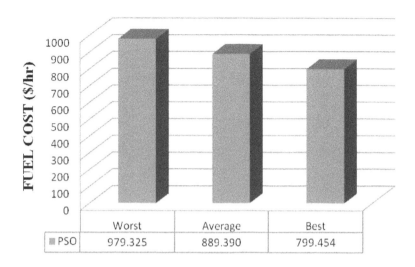

FIGURE 3.8   Fuel cost optimization for 50 trails.

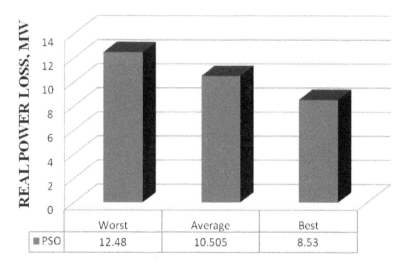

FIGURE 3.9   Real power loss optimization for 50 trails.

3.6 CONCLUSION

This chapter discussed the applications of PSO search algorithm to solve OPF problems. The result of PSO is compared with those of MSFLA, GABC, and BBO algorithms. The obtained simulation results for Standard IEEE 30 bus system using PSO algorithm show reduction in generating cost and losses. In IEEE 30 bus system, the fuel cost is $799.454/hour and the loss is 8.53 MW. By using PSO algorithm, the fuel cost is reduced by 1% and real power loss is reduced by 12.05%.

Finally, by comparing the best-obtained outcomes, it can be concluded that:

The PSO algorithm is very potent and robust for solving the OPF problem, which has been proved several times in various circumstances.

Contrary to other alternatives, the obtained execution times from the proposed approach shows that the PSO algorithm managed to converge to an optimal solution in less execution time than the other alternatives.

3.7 SUMMARY

i. Specific objectives of OPF were elaborated.

ii. Steps involved in solving OPF by PSO approach is discussed with flowchart.

iii. Reduction in fuel cost by 1% and reduction in loss by 12.05% display the supremacy of PSO over other algorithms.

REFERENCES

Attia, AF, Al-Turki, YA & Abusorrah, AM, 2012, 'Optimal power flow using adapted genetic algorithm with adjusting population size', Electric Power Components and Systems, vol. 40, no. 11, pp. 1285–1299.

Azizipanah-Abarghooee, R, Narimani, MR, Bahmani-Firouzi, B & Niknam, T, 2014, 'Modified shuffled frog leaping algorithm for multi-objective optimal power flow with FACTS devices', Journal of Intelligent & Fuzzy Systems, vol. 26, no. 2, pp. 681–692.

Bakirtzis, AG, Biskas, PN, Zoumas, CE & Petridis, V, 2002, 'Optimal power flow by enhanced genetic algorithm', IEEE Transactions on Power Systems, vol. 17, no. 2, pp. 229–236.

Jordehi, AR, 2015, 'Brainstorm optimisation algorithm (BSOA): an efficient algorithm for finding optimal location and setting of FACTS devices in electric power systems', International Journal of Electrical Power & Energy Systems, vol. 69, pp. 48–57.

Kennedy J & Eberhart R, 1995, 'Particle swarm optimization', Proceedings of ICNN'95: International Conference on Neural Networks, IEEE, Australia, vol. 4, pp. 1942–1948.

Malik, IM & Srinivasan D, 2010, Optimum Power Flow Using Flexible Genetic Algorithm Model in Practical Power Systems, IEEE, pp. 1146–1151.

Prathiba, R, Moses, MB & Sakthivel, S, 2014, 'Flower pollination algorithm applied for different economic load dispatch problems', International Journal of Engineering and Technology, vol. 6, no. 2, pp. 1009–1016.

Yan, H & Li, X, 2010, 'Stochastic optimal power flow based improved differential evolution', Proceedings of IEEE Second Conference WRI Global Congress on Intelligent Systems, vol. 3, pp. 243–246.

4 Cuckoo Search Algorithm Based OPF

LEARNING OUTCOME

i. To study the application of cuckoo search algorithm in optimal power flow.
ii. To implement the fuzzy logic controller for determining the reactive power limits.
iii. To make a comparative study on fuel cost with various optimization algorithms.
iv. Comparative analysis of real power losses with different algorithms.

4.1 INTRODUCTION

The invention of flexible alternating current transmission system (FACTS) devices and deregulation of a power sector makes the power system cumbersome and hence the optimal power flow (OPF) has become complex. Recently, metaheuristic algorithms have found major applications in case of complex problems. The cuckoo search (CS) algorithm was first developed by Deb and Yang (2009). CS is a new evolutionary optimization algorithm inspired by the lifestyle of the cuckoo bird family. Cuckoos are well-known birds, not only for their beautiful call, but also because of their aggressive reproduction strategy, by which cuckoos lay their eggs in the nests of other host birds or species. These birds search the best bird's nest to lay their eggs, which gives best breeding. The cuckoo is perhaps the most familiar brood parasite. Some host birds will engage directly with the intruding cuckoo. If the host bird identifies eggs that are not their eggs, then it will either throw those eggs away from its nest or simply destroy its nest and build a new one. In a nest, each egg represents a solution, and cuckoo egg represents a new and good solution. The obtained solution is a new solution based on the existing one and the modification of some characteristics. In the simplest form, each nest has one egg of cuckoo; whereas in complicated cases each nest has multiple eggs representing a set of solutions.

Deb and Yang (2010) introduced novel diagnosis technique using CS algorithm to solve engineering design optimization problems with an objective of minimizing both the weight of the spring and the overall fabrication cost. The results that obtained are compared with other metaheuristic algorithms such as genetic algorithm (GA) and particle swarm optimization (PSO), and CS algorithm is found much superior to other algorithms in all the tested problems. Vazquez (2011) applied CS algorithm and determined the efficiency of the spiking neuron in pattern recognition. In addition, a comparison was made between the CS and differential evolution (DE) algorithms. From the tests, the results obtained with the spiking neuron model trained with the CS algorithm were slightly better than the DE algorithm. Choudhary and Purohit (2011) introduced new heuristic approach using CS for generation of test cases. The outcome of the proposed algorithm is used to achieve multiobjectives of GA. Dhivya et al. (2011) developed the cuckoo-based particle approach (CBPA) to achieve energy-efficient wireless sensor networks (WSNs) and multimodal objective functions. Minimization of energy of WSNs and maximization of lifetime are the two objective functions, and their performances were also measured. Perumal et al. (2011) presented a heuristic method for automation of test data generation, using CS with Lévy flights and Tabu search. The outcome of the developed algorithm performs better in comparison with GA. Civicioglu and Besdok (2011) applied various metaheuristic algorithms such as PSO, DE, and artificial bee colony (ABC) and analyzed the output. They found that CS algorithm provides more precise results than the PSO and ABC algorithms. Kaveh and Bakhshpoori (2011) utilized CS algorithm for optimum design of steel frames. The outcome obtained by the CS is better than other algorithms. Kumar and Chakarverty (2011) used CS algorithm in multiobjective reliability optimization and found that CS is more efficient in terms of cost and performance in comparison with other metaheuristic search algorithms. Burnwal and Deb (2012) proposed CS for scheduling optimization of flexible manufacturing system by minimizing the cost and maximizing the machine utilization time. The result obtained is more efficient compared with other evolutionary techniques such as GA. Natarajan et al. (2012) developed an enhanced cuckoo search (ECS) for optimization of bloom filter (BF) in spam filtering. An ECS algorithm is employed to minimize the total membership invalidation cost of the BFs by finding the optimal false positive rates and number of elements stored in every bin. Yildiz (2012) demonstrates that the application of CS algorithm for solving manufacturing optimization problems found to be more effective. Bacanin and Tuba (2012) developed an object-oriented software system that implements a CS algorithm for unconstrained optimization problems. Their implementation shows that algorithm is superior and is applicable for all new problem-solving. Prakash et al. (2012) proposed a cuckoo optimization algorithm for optimal job allocation of resources on each node. It will allocate the job optimally by taking care of the user's deadline requirement with minimal execution time. Mishra et al. (2015) used CS algorithm to solve OPF with wind power injection. For this, IEEE 30 bus and 57 bus systems are considered. Penalty cost for the wind mill power injection is also considered in their work.

The above section provides enough information on CS algorithm, which has been used for different kinds of optimization problems across various categories. The major field considered for CS algorithm is engineering followed by software testing, pattern recognition, networking WSNs, job scheduling, etc. Based on the statistical results obtained, the majority of techniques are based upon standard CS in combination with Lévy flight mechanism.

However, the previous benchmark results have shown that CS outperforms in comparison with other evolutionary algorithms such as GA, PSO, Tabu search, ABC, and DE. The computation time, convergence rate, and cost are also low in comparison with other algorithms. This chapter clearly explains the concept of CS algorithm and implementation of this algorithm for OPF and gives precise results in comparison with other metaheuristic algorithms.

4.2 CUCKOO SEARCH ALGORITHM

The CS algorithm is a metaheuristic optimization algorithm developed recently for solving optimization problems. It falls under a nature-inspired metaheuristic algorithm, which is based on the brood parasitism of some cuckoo species, along with Lévy flights random walks. In all engineering applications, the search strategy adapted by Cuckoo algorithm (Qin et al., 2014) is to adhere to following rules (Yang and Deb, 2010).

i. Each cuckoo bird lays one egg at a time, and deposits it in an unevenly chosen nest.
ii. The best nests with the highest classes of eggs (solutions) will take them over to the next generations.
iii. The number of existing host nests is fixed, and a host can find out an alien egg with a possibility $pa \in [0, 1]$. In this case, to build an entirely new nest in a new place, the host bird can either throw the egg away or abandon the nest to build a completely new nest in a new location (Yang, 2009).

Built on these three rules, the possibility is that the host bird will either throw the egg away or abandon the nest and build a completely new nest. For simplicity, this last assumption can be approximated by the fraction pa of the n nests that are replaced by new nests (with new random solutions).

4.2.1 GENERATING INITIAL CUCKOO HABITAT

For solving an optimization task, it is necessary that the values of problem variables be formed as an array. Here in CS algorithm, this array is termed as "habitat." The terminologies used in GA and PSO are "Chromosome" and "Particle Position", respectively.

The following are the formulas used for generating a cuckoo habitat:

$$Habitat = [x_1,\ x_2, \ldots, xN_{var} \tag{4.1}$$

In a N_{var}-dimensional optimization problem, a habitat is an array of $1 \times N_{var}$ representing current living position of the cuckoo.

$$Profit = f_{pro}(habitat) = f_{ppro}(x_1, x_2, \ldots, xN_{var}) \tag{4.2}$$

The profit of a habitat is obtained by the evaluation of profit function f_{pro} at a habitat of $(x_1, x_2, \ldots, xN_{var})$.

$$Profit = -Cost(habitat) = -f_c(x_1, x_2, \ldots, xN_{var}) \tag{4.3}$$

The algorithm is intentionally to maximize the profit function.

Each cuckoo bird by nature lays from 5 to 20 eggs. The main part about the real cuckoo is the egg-laying radius (ELR), also termed as a maximum distance, in

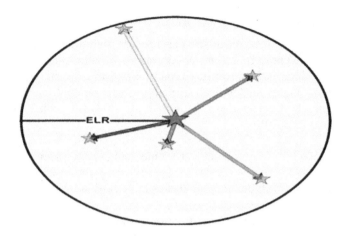

FIGURE 4.1 Random egg laying in ELR. The red star at the center is the initial habitat of the cuckoo with five eggs; pink stars are the eggs' new nest (Rajabioun, 2011).

which cuckoo bird lay eggs within a maximum distance from their habitat, as shown in Figure 4.1.

For an optimization problem, the upper limit is var_{hi} and lower limit is var_{low} for variables; each cuckoo has an ELR that is proportional to the total number of eggs, number of current cuckoo's eggs, and also variable limits of var_{hi} and var_{low}. Hence, ELR is defined as

$$ELR = \alpha \times Number\ of\ current\ cuckoo's\ eggs/total\ number\ of\ eggs \times (var_{hi} - var_{low})$$

Here, α is the integer, supposed to handle the maximum value of *ELR*.

4.2.2 Cuckoos' Style for Egg Laying

Every cuckoo bird starts laying eggs randomly in other host bird's nest within their ELR, as shown in Figure 4.1. Thereafter, the host bird throws some of the non-similar cuckoo eggs out from the nest. Hence, after the egg-laying process, $p\%$ of all eggs (normally 10%), with less profit values, will be killed. Therefore, the eggs will not grow. The rest of the eggs grow in host nests, hatch, and are fed by host birds. Also, only one egg has the chance to grow in a nest. This happens because when a cuckoo egg hatches and the chick comes out, it throws the host bird's own eggs out of the nest. In the case that the host bird's eggs hatch earlier and the cuckoo egg hatches later, the cuckoo's chick eats most of the food the host bird brings to the nest (because of its three-times bigger body, the cuckoo chick pushes other chicks and eats more). After a period of days, the host birds' own chicks die from hunger and only the cuckoo chick remains in the nest.

4.2.3   IMMIGRATION OF CUCKOOS

After cuckoos have laid their eggs, the next stage is the immigration of the cuckoos. The little cuckoos will grow and become mature, and they will reside in their own area and society for some time. When it attains the egg-laying approaches, they immigrate to new and better habitats where their eggs are similar to eggs of the host birds and also where there is more food for new youngsters. This condition will make the cuckoo group form in a different area. The society with the best profit is the goal point for cuckoos to immigrate. As grown-up cuckoos live all over the environment, it is difficult to identify which cuckoo belongs to which group. For providing solutions to this problem, the cuckoo grouping is done with the k-means clustering method. The clustering method means to group cuckoos in a cluster and identify the best group and select the goal habitat. For that, the cuckoo groups are constituted and their mean profit value is calculated. The maximum value of this mean profit values determines the goal group and, consequently, the new destination habitat is the best-group habitat for immigrant cuckoos. When moving toward a goal point, the cuckoos do not fly all the way to the destination habitat. They fly only a part of the way and also have a deviation. This movement of cuckoo bird is clearly displayed in Figure 4.2.

From Figure 4.2, each cuckoo flies only $\lambda\%$ of the total distance toward the goal habitat with a deviation of φ radians. The parameters λ and φ will help the cuckoos search for more positions in all environments. Here $\lambda \sim U(0, 1)$ means that λ is a random number (uniformly distributed) between 0 and 1; ω is a parameter that constrains the deviation from the goal habitat. An ω of $\pi/6$ (rad) seems to be enough for good convergence of the cuckoo population to global maximum profit. When all cuckoos have immigrated toward the goal point and new habitats are specified, each mature cuckoo is given some eggs. Then, considering the number of eggs dedicated to each bird, an *ELR* is calculated for each cuckoo. Afterward, the new egg-laying process restarts.

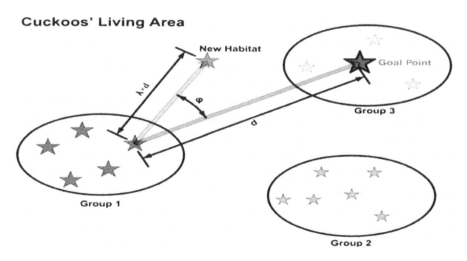

FIGURE 4.2   Immigration of cuckoo (Rajabioun, 2011).

4.2.4 Eliminating Cuckoos in Worst Habitats

There are limits on maximum number of live cuckoos in the cuckoo society. Due to that condition, N_{max} number of cuckoos survive, which have better profit values than the others that demise. This limit is due to food limitations, being killed by predators, and inability to find proper nests for eggs.

4.2.5 Convergence

For the final phase or process of the cuckoo life, the total cuckoo population will migrate to one best habitat with almost maximum similarity of host bird's eggs and with the maximum food resources. This habitat will produce the maximum profit. A few egg losses occur in this best habitat. The flowchart is displayed in Figure 4.3.

The main important issue in CS algorithm is the application of Lévy flights for generating new solutions, $x(t + 1)$:

$$xt + 1 = xt + sEt$$

Here Et is drawn from a standard normal distribution with zero mean and unity standard deviation for random walks, or it is drawn from Lévy distribution for Lévy flights.

4.2.6 Step Size

This random walks can also be linked with the similarity between a cuckoo's egg and the host's egg, which can be tedious in implementation. Then, the step size s determines how far a random walker can go for a fixed number of iterations. Under certain cases, the step size s is too large, the new solution generated will be too far away from the old solution. Such move is unlikely to be accepted. In some cases, if s is too small, the change is too small to be significant, and thereby such search is not effective. Therefore, a proper step size is important for maintaining the best efficient search possible.

4.3 FUZZY LOGIC CONTROLLER

The fuzzy logic is used to find the reactive power to be delivered by the generator, which is very important for CS algorithm to determine optimal real power generation in order to get minimum losses. Fuzzy logic is a multivalued logic system. Fuzzy logic is one of the successful applications of fuzzy set in which the variables are linguistic rather than the numeric variables. Linguistic variables are those variables whose values are sentences in a natural language (such as large or small), represented by fuzzy sets.

Fuzzy set is an extension of a "crisp" set where an element can only belong to a set (full membership) or does not belong at all (no membership). Fuzzy logic control provides an organized way to incorporate human experience in the controller. This fuzzy logic provides a strong framework for achieving robust and simple solutions among different approaches of intelligent computation.

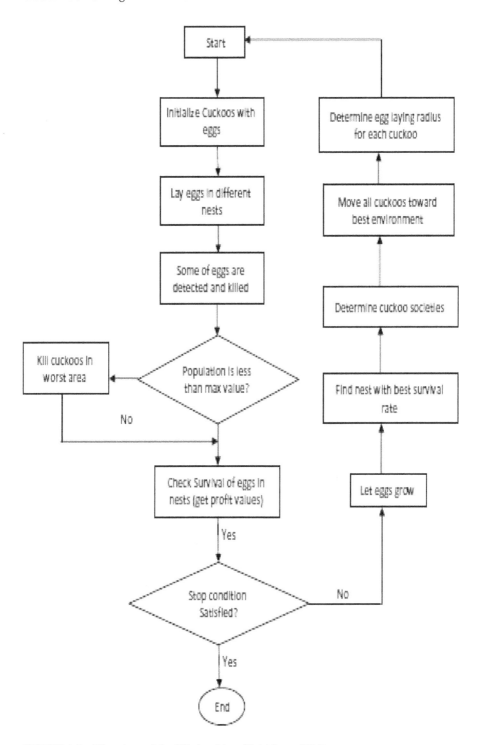

FIGURE 4.3   Flowchart of the CS algorithm (Rajabioun, 2011).

The fuzzy model is a collection of if-then rules with imprecise predicates that use a fuzzy reasoning such as Sugeno and Mamdani models. The Sugeno fuzzy systems can be used to model any inference system in which the output membership functions are either linear or constant, whereas Mamdani type produces either linear or nonlinear output. The fuzzy logic controller consists of four stages: fuzzification of inputs, derivation of rules, inference mechanism, and defuzzification.

Fuzzy logic systems are universal function approximators. In general, the aim of the fuzzy logic system is to yield a set of outputs for given inputs in a nonlinear system, without using any mathematical model, but by using linguistic rules.

The main features of fuzzy logic are as follows:

- Fuzzy logic is conceptually easy to understand. The mathematical concepts behind fuzzy reasoning are very simple. The "naturalness" of fuzzy approach and not its far-reaching complexity makes it better to understand easily.
- Fuzzy logic is flexible. With any given system, it is easy to massage it or layer more functionality on top of it without starting again from scratch.
- Fuzzy logic is tolerant of imprecise data. Everything is imprecise if you look closely enough, but more than that, most things are imprecise even on careful inspection. Fuzzy reasoning builds this understanding into the process rather than tackling it onto the end.
- Fuzzy logic can model nonlinear functions of arbitrary complexity. One can create a fuzzy system to match any set of input-output data. This process is made particularly easy by adaptive techniques like adaptive neuro-fuzzy inference systems (ANFIS), which are available in the Fuzzy Logic Toolbox.
- Fuzzy logic can be built on the experience of experts. In direct contrast to neural networks, which take training data and generate opaque, impenetrable models, fuzzy logic lets one to rely on the experience of people who already understand one's system.
- Fuzzy logic can be blended with conventional control techniques. Fuzzy systems do not necessarily replace conventional control methods. In many cases, fuzzy systems augment them and simplify their implementation.

4.3.1 Fuzzy Set Theory

Zadeh introduced the concept of fuzzy set theory. Recently, the fuzzy set theory applications have received increasing attention in designing intelligent controllers to solve complex problems. Real-world solutions are very often not crisp; rather they are vague, uncertain, and imprecise. Fuzzy logic provides us not only with meaningful and powerful representation for measurement of uncertainties but also with a meaningful representation of vague concepts in natural language. The closer one looks at a real-world problem, the fuzzier becomes its solution. Fuzzy systems can focus on modeling problem characterized by imprecise or ambiguous information. The underlying power of fuzzy set theory is that it uses linguistic variables rather than quantitative variables to represent imprecise concepts (Dombi, 1990). The incorporation of fuzzy set theory and fuzzy logic into computer models has shown tremendous pay off in areas where intuition and judgment still play major role in the model.

Fuzziness describes the ambiguity of an event, whereas randomness describes the uncertainty in the occurrence of the event. In the modern view, uncertainty is considered essential to science; it is not only an unavoidable plague, but it has in fact a great utility. A fuzzy set can be defined mathematically by assigning to each possible individual, in the universe of discourse, a value representing its grade of membership in the fuzzy set.

4.3.2 MEMBERSHIP FUNCTIONS

The membership functions play an important role in designing fuzzy systems. The membership functions characterize the fuzziness in a fuzzy set whether the elements in the set are discrete or continuous in a graphical form for eventual use in mathematical formalism of fuzzy set theory. The shape of membership function describes the fuzziness in graphical form. The shape of membership functions is also important in the development of fuzzy system.

The membership functions can be symmetrical or asymmetrical. A uniform representation of membership functions is desirable (Hiyama et al., 1997). The different types of membership functions are (1) triangular, (2) trapezoidal, (3) Gaussian, (4) sigmoid, and (5) polynomial functions. A membership function associated with a given fuzzy set maps an input value to its appropriate membership value.

4.3.3 MAMDANI FUZZY LOGIC INFERENCE SYSTEM

Mamdani-type of fuzzy logic controller contains four main parts, of which two perform transformations, as shown in Figure 4.4. The four parts are as follows:

- Fuzzifier (transformation 1)
- Knowledge base
- Inference engine (fuzzy reasoning)
- Defuzzifier (transformation 2)

4.3.3.1 Fuzzifier

The fuzzifier performs measurement of the input variables (input signals, real variables), scale mapping, and fuzzification (transformation 1). Thus, all the monitoring input signals are scaled, and the measured signals (crisp input quantities that have numerical values) are transformed into fuzzy quantities by the process of fuzzification. This transformation is performed by using membership functions. In a conventional fuzzy logic controller, the number of membership functions and the shapes of these are initially determined by the user. A membership function has a value between 0 and 1, and it indicates the degree of belongingness of a quantity to a fuzzy set. If it is absolutely certain that the quantity belongs to the fuzzy set, then its value

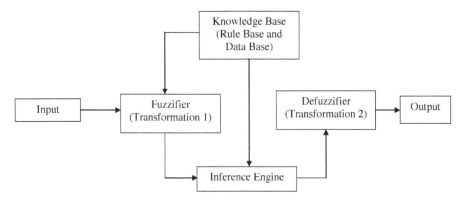

FIGURE 4.4 Mamdani fuzzy logic inference systems.

is 1 (it is 100% certain that the quantity belongs to this set); but if it is absolutely certain that it does not belong to this set, then its value is 0. Similarly, if the quantity belongs to the fuzzy set to an extent of 50%, then the membership function is 0.5.

There are many types of different membership functions: piecewise linear or continuous. The commonly used membership functions are bell-shaped, sigmoid, Gaussian, triangular, and trapezoidal. The choice of the type of membership function used in a specific problem is not unique. Thus, it is reasonable to specify parameterized membership functions, which can be fitted to a practical problem. If the number of elements in the universe X is very large or if a continuum is used for X, then it is useful to have a parameterized membership function, where the parameters are adjusted according to the given problem. Parameterized membership functions play an important role in adaptive fuzzy systems, but are also useful for digital implementation. Due to their simple forms and high computational efficiency, simple membership functions, which contain straight-line segments, are used extensively in various implementations.

4.3.3.2 Knowledge Base

The knowledge base consists of the data base and the linguistic control rule base. The data base provides the information that is used to define the linguistic control rules and the fuzzy data manipulation in the fuzzy logic controller. The rule base contains a set of if-then rules, and these rules specify the control goal actions by means of a set of linguistic control rules. In other words, the rule base contains rules that would be provided by an expert.

The fuzzy logic controller looks at the input signals, and by using the expert rules determines the appropriate output signals (control actions).

4.3.3.3 Inference Engine

It is the kernel of a fuzzy logic controller and has the capability of both simulating human decision-making based on fuzzy concepts and inferring fuzzy control actions by using fuzzy implication and fuzzy logic rules of inference, as shown in Figure 4.3. In other words, once all the monitored input variables are transformed into their

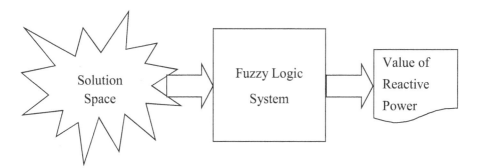

FIGURE 4.5   Fuzzy inference system.

respective linguistic variables, the inference engine evaluates the set of if-then rules and thus result is obtained, which is again a linguistic value for the linguistic variable. This linguistic result has to be then transformed into a crisp output value of the fuzzy logic control.

4.3.3.4   Defuzzification

The output linguistic rule after processing is transformed into output variables/output stabilizing voltage signals. This transformation from output linguistic rule to output variables is called defuzzification.

The value of reactive power generations are calculated using the fuzzy logic system, and it is known as fuzzy inference system as given in Figure 4.5.

4.4   CS ALGORITHM BASED OPF

IEEE 30 bus system is considered for the implementation. In this case, five real power generation, six generator voltage, and four transformer tap positions are used as control variables. To optimize OPF problem, the control variables, real power generation, generator bus voltages, and transformer tap position are considered. The limits on these control variables form prime constraints in addition to power balance condition.

4.4.1   ENCODING

Encoding is the process of converting set of control variables in OPF into optimization problem. Ability of cuckoo is to operate on floating point, and mixed integer makes ease of encoding. The final value of vector gives optimal values of control variables is the optimal solution of OPF. For the evolution and better convergence, fitness function is most important.

4.4.2   FITNESS FUNCTION

An appropriate fitness function is vital for evolution and convergence. Losses are taken as objective, which need to be minimized. Objective function value for a vector is called fitness for the cuckoo egg. Egg represents the solution or the values of the real power generation, which gives the minimum loss. Placing the egg in the nest is equivalent to finding the objective values of the particular set of control variables for the solution, which gives minimum loss in the nest very suitable to lay the egg.

4.4.3 STOPPING CRITERIA

CS improves problems' solution iteration by iteration, and the iteration has to be stopped when either the problem is converged or iteration reached its maximum value. Stopping of iteration is important to provide solution for time complexity. In this research work, maximum number of 200 iterations is considered as stopping criteria.

4.4.4 STEPS INVOLVED IN CS ALGORITHM FOR SOLVING OPF

1. Initialize all the input reactive power limits.
2. Find the fitness of the system using Equation (4.4):

$$fitness = \text{Min} \sum_{k=1}^{N_S} P_{Loss}(k) \qquad (4.4)$$

3. Determine the better fitness and generate the new solution using Equation (4.5):

$$X_I^{t+1} = X_i^t + \alpha \oplus L\acute{e}vy(\lambda) \qquad (4.5)$$

Here, $\alpha > 0$ is the step size, which should be related to the scale of the problem of interest, and the product $\oplus$ means entry-wise multiplications. In this research work, we consider a Lévy flight in which the step-lengths are distributed according to the following probability distribution [Equation (4.6)]:

$$L\acute{e}vy(\lambda) = t^{-\lambda}, 1 < \lambda \le 3 \qquad (4.6)$$

4. Find the fitness probability rate using $pa \in [0, 1]$, whereas the best solution can be determined by the minimum power loss, which is given in the fitness function.
5. Terminate the process.

4.5   COMPUTED RESULTS

IEEE 30 bus system is considered to validate the CS algorithm using MATLAB environment. The IEEE 30 bus system has six generator buses, 24 load buses, and 41 transmission lines. The proposed IEEE 30 system structure is given in Figure 4.6. For the loss minimization, Equation (3.10) is considered as objective function. The system has six generators, four transformers, and forty-one transmission lines. In

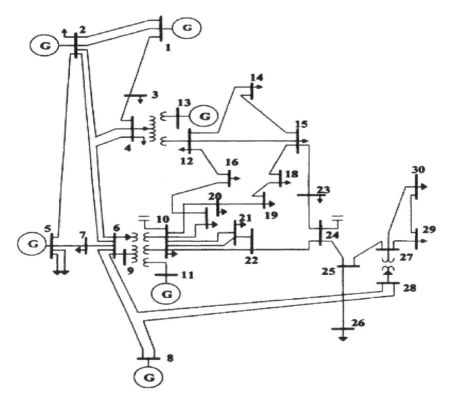

FIGURE 4.6 IEEE 30 bus system single-line diagram. (*Source:* Power Systems Test Case Archive – University of Washington Electrical Engineering.)

this case, five real power generation, six generator voltage, and four transformer tap positions are used as control variables.

The fuzzy logic can produce reactive power generation limits, depending on the expert knowledge rules base. The optimal reactive power limit has been identified using the CS algorithm. This algorithm identifies the minimized power loss of the bus system and the corresponding reactive power limits of the generators. The real power generation, loss, and total generation cost of the system are provided in Table 4.1. From Table 4.1, it is conferred that the CS optimization provides less real power loss compared to other algorithms. The voltage profile of all the bus is given in Figure 4.7. From this plot, it is clear that all the buses' voltage level is more than 0.97.

Figure 4.8 shows the minimization of loss and iterations. The loss is converged at 24th iteration and the loss is reduced to the minimum value of 8.15 MW. The generation cost is converged at 20th iteration at the cost of $799.031/hour. For the generating cost minimization, the quadratic cost function without valve point loading effect and with 100% loading condition is considered. Figure 4.9 shows the convergence curve and time taken to get the convergence is 22 seconds.

The total power generation and demand for different algorithms and the CS method is shown in Figure 4.10. From the figure, it is noted that using CS algorithm,

TABLE 4.1

Optimal Power Generation Given by Cuckoo Search Algorithm

S. No.	Variables	P_{min} (MW)	P_{max} (MW)	MSFLA (2011)	GABC (2015)	BBO (2016)	CS
1	P_{G1} (MW)	50	200	179.1929	177.236	176.611	173.08
2	P_{G2} (MW)	20	80	48.9804	48.701	48.624	47.01
3	P_{G5} (MW)	15	50	20.4517	21.353	21523	22.34
4	P_{G8} (MW)	10	35	20.9264	21.134	21.825	24.25
5	P_{G11} (MW)	10	30	11.5897	11.898	12.170	12.26
6	P_{G13} (MW)	12	40	11.9579	12.00	12.271	12.61
7	Total power generation (MW)	—	—	293.0991	292.322	293.024	291.55
8	Total demand (MW)	—	—	283.4	283.4	283.4	283.4
9	Real power loss (MW)	—	—	9.6991	8.934	9.624	8.15
10	Generating cost ($/hour)	—	—	802.287	800.0963	802.721	799.031

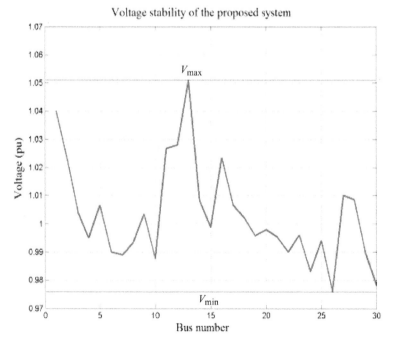

FIGURE 4.7   Voltage level of all the buses.

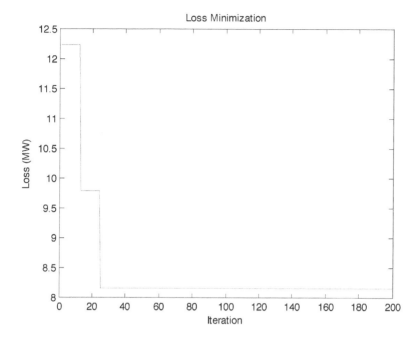

FIGURE 4.8   Real power loss convergence curve.

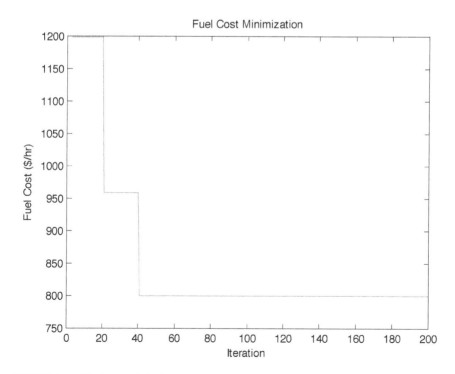

FIGURE 4.9   Fuel cost minimization for the iterations.

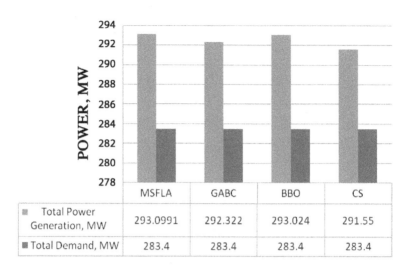

FIGURE 4.10   Comparison of total power generation and demand versus various algorithms.

the total power generation is reduced to 291.55 MW and thereby reduction in losses of the total system. Figure 4.11 shows the real power loss comparison between the existing and CS algorithms; from the figure, it is noted that loss is reduced to 8.15 MW. Figure 4.12 explains about the fuel cost comparison between the already available approach in the literature, and the CS algorithm also shows that fuel cost is reduced to 799.031 $/hour.

In order to validate the effectiveness of algorithm with different initial conditions, the CS algorithm is executed for 50 trails and the best value attained is 799.31, average value is 891.641, and worst value is 984.251, as shown in Figure 4.13. For real power loss again, the CS algorithm is implemented for 50 trails and the best, average, and worst results are shown in Figure 4.14, respectively.

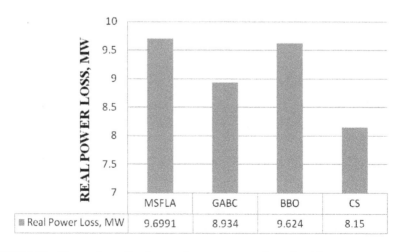

FIGURE 4.11   Comparison of real power losses versus various algorithms.

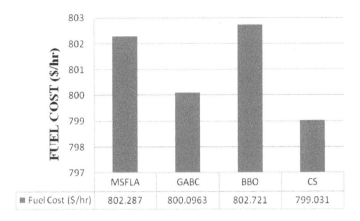

FIGURE 4.12   Comparison of fuel cost versus various algorithms.

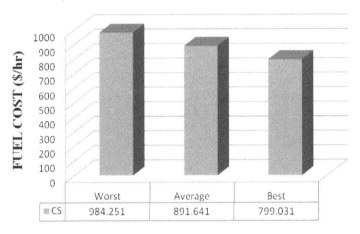

FIGURE 4.13   Fuel cost optimization for 50 trails.

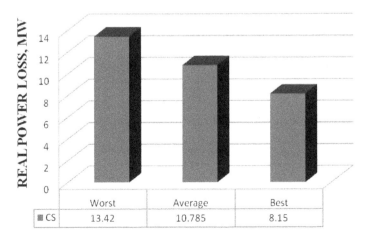

FIGURE 4.14   Real power loss optimization for 50 trails.

4.6 CONCLUSION

This chapter discusses the importance of fuzzy logic and CS optimization algorithm to solve OPF problems. The application of fuzzy logic offers the best limits for the reactive power of the generators, and the CS algorithm is applied to find the best real power generation for the minimum loss and generating cost without valve point loading effect. The simulation results show reduction in generating cost losses and improvement in voltage profile, thereby providing stability for Standard IEEE 30 bus system. The real power loss is reduced by 8.15 MW, and the fuel cost is reduced by \$799.031/hour. The real power loss is reduced by 15.92%, and the fuel cost is reduced by 1% by implementing this metaheuristic algorithm.

Finally, by comparing the best-obtained outcomes, it can be concluded that:

i. The CS approach provides more robust and effective outcome for solving the OPF problem in case studies that has been proved several times in different circumstances. As a case in point, it managed to reduce the total generation cost, thereby reducing real power loss, which is quite a cogent achievement.

A suitable fuzzy decision-making tactic has been implemented based on the expert knowledge rules base for producing the reactive power generation limits. With respect to other alternatives, the obtained execution times from the proposed approach show that the proposed algorithm managed to converge to an optimal solution in less execution time than the other alternatives of the defined case studies.

4.7 SUMMARY

The benchmark tests also show that CS outperforms other evolutionary algorithms such as GA, PSO, Tabu search, ABC, and DE. Based on the experimental results, CS also improves the performances and is better than all other algorithms in terms of computation time, convergence rate, and cost.

REFERENCES

Bacanin, N & Tuba, M, 2012, 'Artificial bee colony (ABC) algorithm for constrained optimization improved with genetic operators', Studies in Informatics and Control, vol. 21, no. 2, pp. 137–146.

Burnwal, S & Deb, S, 2012, 'Scheduling optimization of flexible manufacturing system using cuckoo search-based approach', International Journal of Advanced Manufacturing Technology, vol. 64, pp. 951–959.

Choudhary, K & Purohit, GN, 2011, 'A new testing approach using cuckoo search to achieve multi-objective genetic algorithm', Journal of Computing, vol. 3, no. 4, pp. 117–119.

Civicioglu, P & Besdok, E, 2011, 'A conceptual comparison of the cuckoo-search, particle swarm optimization, differential evolution and artificial bee colony algorithms', Artificial Intelligence Review, vol. 39, pp. 315–346.

Deb, S & Yang, X-S, 2009, 'Cuckoo search via levy flights', 2009 World Congress on Nature & Bilogically Inspired Computing (NaBIC).

Deb, S & Yang, X-S, 2010, 'Engineering optimisation by cuckoo search', International Journal of Mathematical Modelling and Numerical Optimisation", vol. 1, no. 4, p. 330.

Dombi, J 1990 'Membership function as an evaluation', Fuzzy Sets and Systems, vol.35, no.1, pp. 1–21.

Hiyama, T, Ueki, Y & Andou, H, 1997, 'Integrated fuzzy logic generator controller for stability enhancement', 1997, IEEE Transactions on Energy Conversion, vol. 12, no. 4, pp. 400–406.

Kaveh, A & Bakhshpoori, T, 2011, 'Optimum design of steel frames using cuckoo search algorithm with Lévy flights', The Structural Design of Tall and Special Buildings Struct, Wiley, Vol. 22, no. 13.

Kumar A & Chakarverty, S, 2011, 'Design optimization for reliable embedded system using cuckoo search', 3rd International Conference on Electronics Computer Technology (ICECT) USA, IEEE, vol. 1, pp. 264–268.

Manian, D, Sundarambal, M & Nithissh Anand, L, 2011, 'Energy efficient computation of data fusion in wireless sensor networks using cuckoo based particle approach (CBPA)', International Journal of Communications, Network and System Sciences, vol. 4, no. 04, pp. 249–255.

Mishra, C, Singh, SP & Rokadia, J, 2015, 'Optimal power flow in the presence of wind power using modified cuckoo search', IET Generation, Transmission and Distribution, vol. 9, no.7, pp. 615–626.

Mohan, P, Saranya, R, Rukmani Jothi, K & Vigneshwaran, A, 2012, 'An optimal job scheduling in grid using cuckoo algorithm', International Journal of Computer Science and Telecommunications, vol. 3, no. 2, pp. 65–69.

Natarajan, A, Subramanian, S & Premalatha, K, 2012, 'A comparative study of cuckoo search and bat algorithm for bloom filter optimization in spam filtering', International Journal of Bio-Inspired Computation, vol. 4, no. 2, pp. 89–99.

Perumal, K, Ungati, JM, Kumar, G, Jain, N, Gaurav, R & Srivastava, PR, 2011, 'Test data generation: A hybrid approach using cuckoo and tabu search', SEMCCO 2011, Part II, LNCS 7077, pp. 46–54.

Qin, S, Liu, F, Wang, J & Sun, B, 2014, 'Analysis and forecasting of the particulate matter (PM) concentration levels over four major cities of China using hybrid models', Atmospheric Environment, vol. 98, pp. 665–675.

Rajabioun, R, 2011, 'Cuckoo optimization algorithm', Applied Soft Computing, vol. 11, no. 8, pp. 5508–5518.

Vazquez, RA, 2011, 'Training spiking neural models using cuckoo search algorithm', IEEE Congress of Evolutionary Computation (CEC), pp. 679–686.

Yildiz, AR, 2012, 'Cuckoo search algorithm for the selection of optimal machining parameters in milling operations', International Journal of Advanced Manufacturing Technology, vol. 64, pp. 1–4.

5 Firefly Algorithm Based OPF

LEARNING OUTCOME

i. To study about the basic principles of firefly algorithm (FFA) and its applications.
ii. To learn about steps involved in the execution of FFA in optimal power flow problem.
iii. Comparative study between different algorithms.

5.1 INTRODUCTION

In the last two decades, more than 30 metaheuristic algorithms such as differential evolution (DE), genetic algorithm (GA), particle swarm intelligence (PSO), bat algorithm, firefly algorithm (FFA), and cuckoo search algorithm have been introduced that have shown potential in solving complex engineering problems. These algorithms are developed by mimicking the scenario from nature; for example, the GA algorithm based on Darwin theory of survival of the fittest; PSO algorithm based on the pattern in which a swarm moves following each other. Among these new algorithms, FFA has been found more efficient in dealing with multimodal, global optimization problems Khan et al. (2016). The above-mentioned intelligent algorithms have been adapted to solve many power systems problems, in particular the optimal power flow (OPF) problem. The OPF problem may have multiobjectives and become nonlinear and constrained optimization problem. Multiobjective optimization is a process of solving a problem by simultaneously optimizing two or more objectives subjected to constraints. Bakirtzis et al. (2002) used enhanced genetic algorithm (EGA) to solve OPF. The objective function of OPF is converted into fitness function with little modification, as GA is suitable for maximization, whereas OPF is a minimization problem. Irfan Mulyawan Malik and Dipti Srinivasan (2010) used GA to solve OPF for which they considered elitism and nonuniform mutation rate. Vaisakh and Srinivas (2008) used DE algorithm to solve OPF. Generator real power except slack bus, generator voltage magnitudes, and transformer tap settings are considered as control variables and converted into vector of DE algorithm.

Apostolopoulos and Vlachos (2011) used FFA for solving economic emission load dispatch (EELD) problem. Yang (2012) used FFA for solving nonconvex valve point loaded economic dispatch problem. Khadwilard et al. (2011) developed FFA for solving the job shop scheduling problem (JSSP). Senthilnath et al. applied FFA for clustering data objects into groups according to the values of their attributes. The performance of FFA for clustering was compared with the results of other nature-inspired algorithms like artificial bees colony (ABC) and PSO. The performance measure used in the comparison was the classification error percentage (CEP) that is defined as a ratio of the number of misclassified samples in the test data set to the total number of samples in the test data set. The authors concluded from the obtained results that FFA was the efficient method for clustering. FFA is a computationally efficient, nature-inspired, population-based metaheuristic that derives its solution approach based upon the characteristics of fireflies. The following are the advantages of FFA:

i. FFA can automatically subdivide its population into subgroups, due to the fact that local attraction is stronger than long-distance attraction. As a result, FFA can deal with highly nonlinear, multimodal optimization problems naturally and efficiently.

ii. FFA does not use historical individual best sn_i, and there is no explicit global best gn either. This avoids any potential drawbacks of premature convergence as those in PSO. In addition, FFA does not use velocities, and there is no problem as that associated with velocity in PSO.

iii. FFA has an ability to control its modality and adapt to problem landscape by controlling its scaling parameter such as γ. In fact, FFA is a generalization of simulated annealing (SA), PSO, and DE, as seen clearly in the next section. In this chapter, FFA is used for multiobjective optimization along with unified power flow controller (UPFC) for OPF. This algorithm is an innovative optimization algorithm that is used for optimizing the objective function (Figure 5.1).

5.2   FFA

Fireflies are among the most charming of all insects. They live in warm environments and are found to be most active in summer nights. Fireflies are characterized by their flashing light produced by biochemical process bioluminescence. Those flashing light may serve as the primary courtship signals for mating. The flashing light may also be used to warn off potential predators apart from mating. Some adult fireflies are not capable of producing bioluminescence, and so they attract their mates by pheromone, like ants.

The lantern is the light-producing organ present in the fireflies, and the lanterns' light production is initialized by signals originating within the central nervous system of firefly. Majority of firefly species rely on bioluminescent courtship signals. Typically, the first signalers are flying males, who try to attract flightless females on the ground. In response to these signals, the females emit continuous or flashing lights. Both mating partners produce distinct flash signal patterns that are precisely timed in order to encode information like species identity and sex. Females are attracted according to behavioral differences in the courtship signal. Typically, females prefer brighter male flashes. It is well known that the flash intensity varies with the distance from the source. Fortunately, in some firefly species, females cannot discriminate between more distant flashes produced by stronger light sources and closer flashes produced by weaker light sources. Firefly flash signals are highly conspicuous and may therefore deter a wide variety of potential predators. In the

FIGURE 5.1   Taxonomy of firefly algorithm (Fister et al., 2013).

sense of natural selection where only the best adapted individual can survive, flash signals evolve as defense mechanisms that serve to warn potential predators.

For optimization, flashing light is formulated based on objective function. Brightest firefly is the most optimal solution for the problem under consideration. A firefly is a set of control variables of the problem being considered. Brightness of the firefly is calculated by evaluating the objective function to be optimized. This algorithm may be used for maximization or minimization problems. FFA has idealization as compared to the natural firefly:

- Firefly is unisex and attracted by another firefly in spite of sex.
- Firefly moves toward brightest; if there is no brighter one, then the firefly moves randomly in solution space.
- Brightness of firefly is affected by problem nature.

General form FFA optimization is a maximization of objective function subjected to constraints. FFA moves fireflies toward global optimal solution spot through iteration by iteration. A firefly is a set of control variables and its light intensity is objective function or fitness value of the firefly. The process of FFA is create or initialize fireflies, find brightness of firefly, move each firefly toward the brightest one, and find global brightest to give optimum value.

The algorithm undertakes that all fireflies are unisex, indicating that any firefly can be attracted by any other firefly. The attractiveness of a firefly is directly proportional to its brightness, which depends upon the objective function. A firefly will be attracted to a brighter firefly. Moreover, the brightness decreases with distance according to the inverse square law, as given in Equation (5.1):

$$I \propto 1/r \tag{5.1}$$

where
  I is the light intensity
  r is the distance

When a light passes through a medium with a light absorption coefficient γ, then the light intensity is given by Equation (5.2):

$$I = Ioe^{-\gamma r^2} \tag{5.2}$$

where I_0 is the light intensity at the source.

The brightness β of fireflies is proportional to their light intensities $I(r)$. Hence, brightness can be given by Equation (5.3):

$$\beta = \beta_0 e^{-\gamma r^2} \tag{5.3}$$

A generalized brightness function for $\omega \geq 1$ is given in Equation (5.4). In fact, any monotonically decreasing function can be used.

$$\beta = \beta_0 e^{-\gamma d^\omega} \tag{5.4}$$

In the algorithm, a randomly generated feasible solution, termed fireflies, will be allotted with a light intensity based on their performance in the objective function.

This light intensity will be used to determine the brightness of the firefly, since it is directly proportional to its light intensity. For minimization problems, a solution with smallest functional value will be assigned with highest light intensity. Once the intensity or brightness of the solutions is assigned, each firefly will follow fireflies with better light intensity. For the brightest firefly, it will perform a local search by randomly moving in its neighborhood. Hence, for two fireflies, if firefly j is brighter than firefly i, then firefly i will move toward firefly j using the updating formula given in Equation (5.5):

$$x_i := x_i + \underbrace{\beta_0 e^{-\gamma r_{ij}^2}}_{\tilde{\beta}}(x_j - x_i) + \alpha(\varepsilon() - 0.5) \tag{5.5}$$

where β_0 is the attractiveness of x_j at $r = 0$. Yeomans (2018) recommended that $\beta_0 = 1$ for implementation, γ is an algorithm parameter that determines the degree to which the updating process depends on the distance between the two fireflies, α is the algorithm parameter for the step length of the random movement, and $\varepsilon()$ is a random vector from uniform distribution with values between 0 and 1. For the brightest firefly, x_b, the second expression in Equation (5.5) can be neglected, as given in Equation (5.6):

$$x_b := x_b + \alpha(\varepsilon() - 0.5) \tag{5.6}$$

These updates of the location of fireflies continue with iteration until a termination criterion is met. The flowchart for the FFA is shown in Figure 5.2.

5.3 FFA-BASED OPF

To optimize the OPF problem, the control variables considered are real power generation, generator bus voltages, and transformer tap position. The limits on these control variables form the prime constraints in addition to power balance condition. The real values of these control variables are used to form a firefly. These fireflies form population and initialized randomly from the solution space and then evolution is carried out using its brightness and distance from the brightest firefly.

5.3.1 ENCODING

Encoding is the process of converting a set of control variables in OPF into firefly for optimization. The ability of FFA is to operate on floating point, and the mixed integer makes ease of encoding. The final iteration of FFA gives the global bright firefly, which is the optimal or best solution of OPF. For the evolution and better convergence, the fitness function is most important.

5.3.2 FITNESS FUNCTION

An appropriate fitness function (brightness) is vital for the evolution and convergence of FFA. It is an OPF objective function and penalty function if any. FFA evaluates brightness of each firefly in the population. Objective function value for a firefly is called brightness of the firefly. FFA makes a firefly to move toward brighter firefly in the population. The distance moved and the brightness of each firefly are calculated, and the best firefly (global best) is calculated in the iteration. Improvement in solution is achieved iteration by iteration and final iteration provides global best optimal solution to OPF.

5.3.3 ATTRACTIVENESS

Firefly moves toward more attractiveness. This attractiveness of the considered firefly along with others is calculated using the function. This attractiveness decreases with increase in the distance between fireflies. The main reason for reduction in

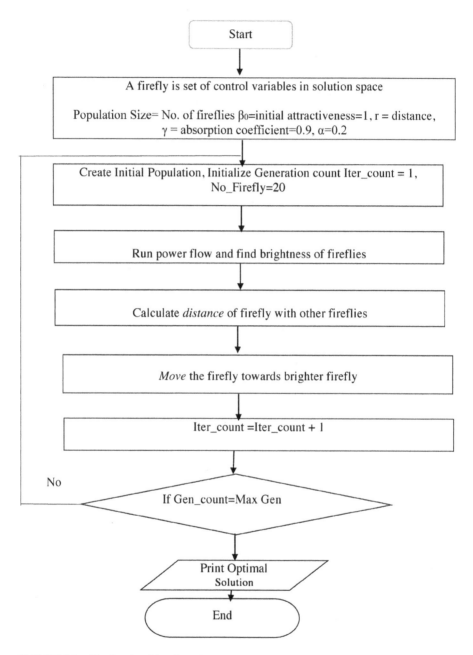

FIGURE 5.2   Firefly algorithm flowchart.

attractiveness is that absorption factors in nature are implemented by using absorption coefficient. This function is monotonically decreasing function as given in Equation (5.7):

$$\beta = \beta_0 \exp(-\gamma\, r_2) \tag{5.7}$$

where

β is the attractiveness of a firefly

β_0 is the initial attractiveness

γ is the absorption coefficient

r is the distance between fireflies

5.3.4 DISTANCE

The distance between fireflies i and j is calculated using Cartesian distance as given in Equation (5.8):

$$r_{ij} = \|x_i - x_j\| \sqrt{\sum_{k-1}^{d} (x_{i,k} - x_{j,k})^2} \tag{5.8}$$

In a two-dimensional solution space, the distance between i and j fireflies may be calculated as given in Equation (5.9):

$$r_{ij} = \sqrt{(x_i - x_j)^2 + (y_i - y_j)^2} \tag{5.9}$$

5.3.5 MOVEMENT

Movement of the ith firefly toward the jth brighter firefly is based on attractiveness and the distance between them, as given in Equation (5.10).

$$x_i^{k+1} = x_i^k + \beta_0 * \exp(-\gamma r_2) * (x_{jk} - x_{ik}) + \alpha * \epsilon_{ik} \tag{5.10}$$

where the left side first term is initial position of the ith firefly, the second term gives attractiveness toward jth firefly, and the third term introduces random movement in the ith firefly. Initial attractiveness β_0 is taken as 1.0; absorption coefficient γ is taken as 0.9. Randomizing coefficient α rang in between 0 and 1, in this work it is taken as 0.2; ε_i is the randomization vector that ranges from 0 to 0.5.

5.3.6 STOPPING CRITERIA

Fireflies moves randomly and try to attract toward brighter firefly. FFA improves problems' solution iteration by iteration, and the iteration has to be stopped when either the problem is converged or the iteration has reached its maximum value. Stopping of iteration is important to provide solution for time complexity. In this research work, maximum number of 200 iterations is considered as stopping criteria.

5.4 STEPS INVOLVED IN FFA FOR SOLVING OPF

The following are the steps involved in FFA for solving OPF:

1. Ensure firefly is a set of control variables in OPF and UPFC.
2. Initialize fireflies in the population within solution space.

3. Use OPF objective function to find brightness of firefly.
4. Calculate attractiveness of firefly with other fireflies.
5. Calculate distance between fireflies.
6. Move firefly i toward firefly j using Equation (5.7).
7. Rank the fireflies and find the current global best.
8. Repeat steps 4–7 till stopping criterion is satisfied.
9. Print the optimal result after stopping criterion is satisfied.

5.5 CASE STUDY AND DISCUSSION

In order to validate the algorithm, the test case IEEE 30 bus is considered. The various parameters that are considered to evaluate the performance are reactive power, voltage, and losses. Dedicated MATLAB software is used to develop and implement the program. The IEEE 30 bus system has 6 generator buses, 24 load buses, and 41 transmission lines.

For UPFC, three control variables are included for the position, shunt, and series injection. The optimal real power loss has been identified using FFA. FFA identifies the minimized power loss of the bus system and the corresponding reactive power limits of the generators. The proposed system also analyzes the voltage stability of the system, which is given in the following. The proposed IEEE 30 system structure is given in Figure 5.3.

For the minimization of generating cost and power loss for the economical operation of power system, the cost of generating unit has to be minimized as far as possible. This obeys the quadratic cost function. The coefficient of the cost function is given in Table 5.1. Real power generation limits of generators are also given in Table 5.1.

Transmission line real power loss minimization is the major component of reactive power optimization and it needs more attention. This case considers that only the real power loss minimization, voltage improvement, and loss minimization lead to minimum generating cost. The problem is solved in the baseline scenario; it is optimized using FFA with UPFC included in the system to get a better optimized result. The optimal allocation of UPFC in buses and lines are represented in Table 5.2. In this case, FFA better optimizes both real power loss and fuel cost, as given in Table 5.3. The table shows the comparison results between the existing and proposed method results. From the results, it is clear that UPFC placement gives best OPF. The reduction in loss indicated by FFA algorithm is highly encouraging: it is only 4.65 MW. Another important objective of the proposed method is fuel cost and it is also reduced to $803.15/hour.

The UPFC connection between the two buses in the power system is identified by the FFA. The optimized values of location and power injection are given in Table 5.2.

Figure 5.4 shows the convergence curve for the loss. From the curve, it is found that the loss converged on 8.72 MW at 54th iteration. The generation cost converged on $786.632/hour at 47th iteration, as shown in Figure 5.5. The time taken to attain the convergence is 29 seconds.

Figure 5.6 shows the total power generated and demand for DE algorithm and FFA with STATCOM FACTS devices and FFA with UPFC. From the figure, it is noted that the real power generated with UPFC is reduced to 292.12 MW, thereby reducing the losses in the system. The real power losses in DE with STATCOM, FFA

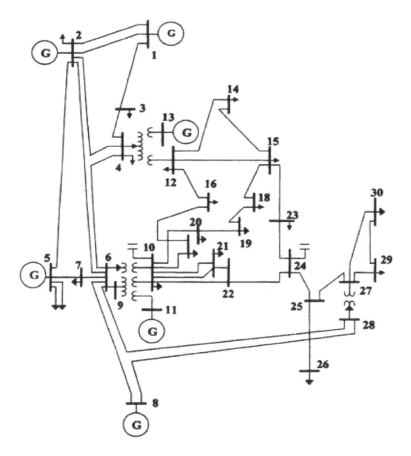

FIGURE 5.3   IEEE system architecture.

TABLE 5.1
Cost Coefficients of Generators

S. No.	Bus No.	Min Real Power (MW)	Max Real Power (MW)	Alpha ($/hour)	Beta ($/MWhr)	Gamma ($/MW²hr)
1	1	50	200	0.00375	2	0
2	2	20	80	0.0175	1.75	0
3	5	15	50	0.0625	1	0
4	8	10	35	0.0083	3.25	0
5	11	10	30	0.025	3	0
6	13	12	40	0.025	3	0

TABLE 5.2

UPFC Location and Power Injection Value

S. No.	Sending-end Bus	Receiving-end Bus	Series MW	Shunt MVAr
1	6	7	4.23	25.2553

with STATCOM, and FFA with UPFC are compared and shown in Figure 5.7. From the figure it is found that the result FFA with UPFC produces less loss in comparison with other devices. The generation cost comparison with DE with STATCOM, FFA with STATCOM, and FFA with UPFC are shown in Figure 5.8. It is proven that FFA with UPFC provides better result in comparison with algorithms.

To take care of different initial conditions, the algorithm is executed for 50 trails and the worst, average, and best results are shown in Figures 5.9 and 5.10. Fuel cost and real power loss for IEEE 30 bus system are optimized with UPFC using FFA. Figures 5.9 and 5.10 show the fuel cost optimization for 50 trails and real power loss optimization for 50 trails, respectively.

TABLE 5.3

Comparison of Objective Terms in FFA

S. No.	Variables	P_{min} (MW)	P_{max} (MW)	Chandrasekar and Rajasekar (2015)	Ponnin Thilagar and Harikrishnan (2015)	FFA-UPFC
1	P_{G1} (MW)	50	200	192.565	151.314	172.13
2	P_{G2} (MW)	20	80	48.7254	42.5989	45.33
3	P_{G5} (MW)	15	50	19.7473	24.0831	21.19
4	P_{G8} (MW)	10	35	11.5553	31.9497	21.32
5	P_{G11} (MW)	10	30	10	24.9827	15.63
6	P_{G13} (MW)	12	40	12	19.5219	12.29
7	Device location	—	—	12	15	6–7
8	Device size	—	—	51.0154 MVAR	86.0452 MVAR	4.23 MW 25.2553 MVAR
9	Total power generation (MW)	—	—	296.687	294.45	292.12
10	Total demand (MW)	—	—	283.4	283.4	283.4
11	Real power loss (MW)	—	—	11.1929	11.05	8.72
12	Generation cost ($/hour)	—	—	805.889	826.120	786.632

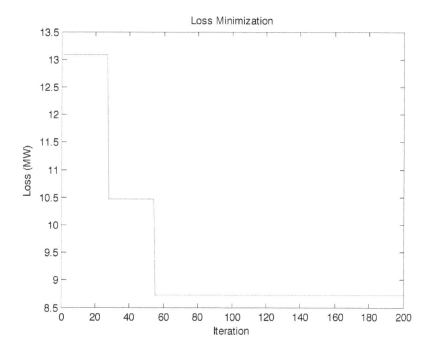

FIGURE 5.4   Real power loss convergence curve.

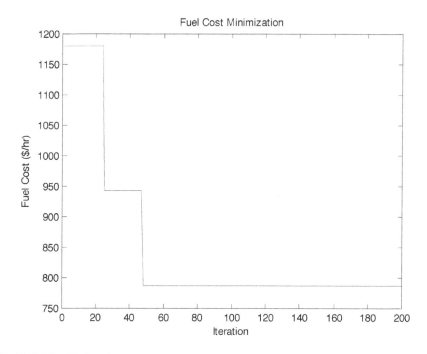

FIGURE 5.5   Fuel cost convergence curve.

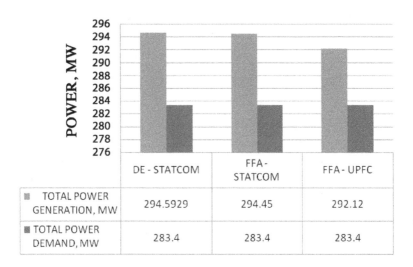

FIGURE 5.6   Comparison of total power generation and demand versus various algorithms.

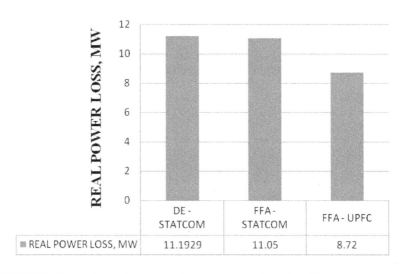

FIGURE 5.7   Comparison of real power losses versus various algorithms.

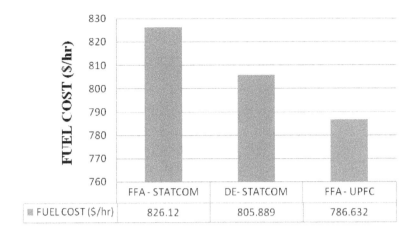

FIGURE 5.8   Comparison of fuel cost versus various algorithms.

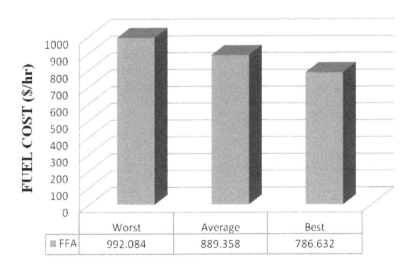

FIGURE 5.9   Fuel cost optimization for 50 trails.

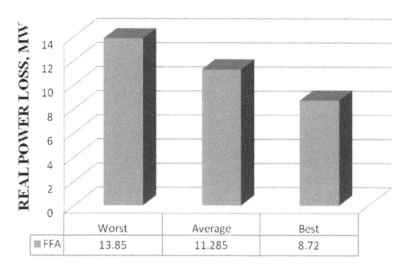

FIGURE 5.10   Real power loss optimization for 50 trails.

5.6   CONCLUSION

In this chapter, the importance of FFA was narrated and applied to solve complex OPF problem with UPFC. FFA use OPF objective function to find the brightness of fireflies. Attractiveness of firefly depends on distance and the firefly moves toward brighter firefly. The obtained simulation results show reduction in generating cost and losses for Standard IEEE 30 bus system. Also, it gives the optimal location of UPFC and the amount of reactive power injection from UPFC. Due to UPFC connection, the fuel cost is reduced by 2.38% and losses are reduced by 21.08%.

5.7   SUMMARY

   i. Better understanding of FFA with basic concepts and applications to solve the OPF problem.
   ii. FFA with different FACTS devices is analyzed and simulation results are presented.
   iii. FFA with UPFC provides better optimal result, with fuel cost reduced by 2.5% and losses by 21.08%.

REFERENCES

Apostolopoulos, T & Vlachos, A, 2011, 'Application of the firefly algorithm for solving the economic emissions load dispatch problem', International Journal of Combinatorics, vol. 2011, pp.1–23.

Babu, TG & Srinivas, GN, 2017, 'Enhancement of ATC with FACTS device using firefly algorithm', International Journal of Applied Engineering Research, vol. 12, no. 20, pp. 10269–10275.

Bakirtzis, AG, Biskas, PN, Zoumas, CE & Petridis, V, 2002, 'Optimal power flow by enhanced genetic algorithm', IEEE Transactions on Power Systems, vol. 17, pp. 229–236.

Chandrasekar, A, & Rajasekar, P, 2015, 'Solving optimal power flow with facts device using de algorithm', ARPN Journal of Engineering and Applied Sciences, vol. 10, no. 2, pp. 933–939.

Fister, I, Fister, I, Jr, Yang, X-S & Brest, J, 2013, 'A comprehensive review of firefly algorithms', Swarm and Evolutionary Computation, vol. 13, pp. 34–36.

Johari, NF, Zain, AM, Mustaffa, N & Udin, A, 2013, 'Firefly algorithm for optimization problem', Applied Mechanics and Materials, vol. 421, pp. 512–517.

Khadwilard, A, Chansombat, S, Thepphakorn, T & Thapatsuwan, P, 2011, 'Investigation of firefly algorithm parameter setting for solving job shop scheduling problems', ORNET2011.

Khan, WA, Hamadneh, NN, Tilahun, SL & Ngnotchouye, JMT, 2016, 'A review and comparative study of firefly algorithm and its modified versions', in Baskan, O, Ed. Optimization Algorithms: Methods and Applications, IntechOpen.

Kumbharana, S, 2013, 'Solving travelling salesman problem using firefly algorithm', International Journal for Research in Science & Advanced Technologies, vol. 2, no. 2, pp. 053–057.

Malik, IM & Srinivasan, D 2010, 'Optimum power flow using flexible genetic algorithm model in practical power systems' 2010 Conference Proceedings IPEC.

Ponnin Thilagar, P & Harikrishnan, R, 2015, 'Application of intelligent firefly algorithm to solve OPF with STATCOM', Indian Journal of Science and Technology, vol. 8, no. 22, pp. 1–5.

Sahu, RK, Panda, S & Padhan, S, 2015, 'A hybrid firefly algorithm and pattern search technique for automatic generation control of multi area power systems', International Journal of Electrical Power & Energy Systems, vol. 64, pp. 9–23.

Senthilnath, J, Omkar, SN & Mani, V, 2011, 'Clustering using firefly algorithm: Performance study,' Swarm and Evolutionary Computation, vol. 1, no. 3, pp. 164–171.

Vaisakh, K & Srinivas, LR 2008, 'Differential evolution based OPF with conventional and non-conventional cost characteristics', Joint International Conference on Power System Technology and IEEE Power India Conference.

Yang, X-S, 2012, 'Firefly algorithm for solving non-convex economic dispatch problems with valve loading effect', Applied Soft Computing, vol. 12, pp. 1180–1186.

Yeomans, JS, 2017, 'An efficient computational procedure for simultaneously generating alternatives to an optimal solution using the firefly algorithm', Part of the Studies in Computational Intelligence Book Series (SCI, vol. 744). Nature-Inspired Algorithms and Applied Optimization, pp. 261–273.

6 Flower Pollination Algorithm Based OPF

LEARNING OUTCOME

i. To provide an overall study of flower pollination algorithm (FPA) along with its types.
ii. To study the performance of FPA approach for solving objective functions: generating cost minimization, emission parameter, and transmission loss.
iii. Comparative outcome of FPA with other different approaches.

6.1 INTRODUCTION

Over the last few decades, many nature-inspired algorithms have been introduced and implemented whenever the world encountered a problem. The conventional optimization does not work for complex problems, particularly with nonlinearity and multimodality. Recently, the application of nature-inspired metaheuristic algorithms to tackle such complex problems has proven that metaheuristic approach works successfully. Despite the diversity of metaheuristic inspirations, it follows up a common algorithmic approach that trade-off between exploration and exploitation. Exploration tends to search space by randomization, whereas exploitation makes the algorithm to search around the current best solution. Metaheuristic approach has been a proven approach toward finding an optimal quality solution if it can provide a balance between the exploitation and exploration of search space (Blum and Roli 2003). The nature-inspired algorithms are based on the characteristics of biological systems. With the advances in genetic algorithms, researchers have proposed several bio-inspired algorithms for optimizing the solution. Genetic algorithms were based on the Darwinian evolution of biological systems, particle swarm optimization (PSO) algorithm was based on the swarm behavior of birds, bat algorithm (BA) was based on the echolocation behavior of microbats, and firefly algorithm (FFA) was based on the flashing light patterns of tropic fireflies.

As around 75% of plants on Earth belong to flowering plants, various authors have paid attention to simulate the evolution of flowers. In 2006, Kazemian et al. developed a clustering algorithm that simulates flowers pollination by bees. Later the flower pollination algorithm (FPA) was proposed by Yang in 2012. This algorithm was inspired by reproduction procedure in plants. The application of FPA gains more importance in solving real-life complex problems.

FPA is used for various applications such as parameter estimation in solar photovoltaic (PV) systems (Kumar and Natarajan, 2017), optimal capacitor locations (Abdelaziz et al., 2016), dynamic multiobjective optimal dispatch for wind-thermal system (Dubey et al. 2015), congestion management (CM) problem (Deb and Goswami 2016).

A comparative analysis was made by Sakib et al. (2014) between FPA and BA on various unimodal and multimodal benchmarks. They found that FPA outperforms BA in terms of convergence, solution reliability, and consistency on continuous optimization problems.

Hegazy et al. (2015) also made a comparative study between the various algorithms, such as modified cuckoo search (MCS) (Walton et al. 2011) and artificial bee colony (ABC) (Karaboga et al. 2005), and found that the results obtained by FPA were superior. Ouadfel and Taleb-Ahmed (2016) conducted comparative study of FPA on thresholding of multilevel image with social spiders optimization (SSO) algorithm (Cuevas & Cienfuegos 2013) and the experimental results proved that the performance of FPA is more suitable for small numbers of thresholds rather than the larger ones. Rathasamuth and Nootyaskool (2016) compared between FPA, PSO, and GA on discrete search space. The convergence curve shown by FPA has better convergence rate than other approaches.

The optimal power flow (OPF) is considered to be an obvious and integral tool in power systems operation and control. The concept was introduced by Carpentier in the early 1960s. The ultimate goal of the OPF is to obtain the most economical combination to precisely serve the total demand of the system without any load shedding or islanding. It should be noted that the aforementioned evolutionary-based optimization algorithms offer many advantages compared to mathematically based methods. However, nature-based algorithms are a double-edged sword, and their benefits and liabilities are inextricably interwoven; consequently, to have an optimal, fast, and reliable solution, they must be handled in such a manner that their merits outweigh their demerits. These methods are not likely to be fast and computationally efficient in most situations, and this is due to the qualitative features of evolutionary processes, which can be a recipe for premature convergence. Furthermore, four objective functions with qualitative discrepancies, such as total generation cost, active power transmission losses, and emission and voltage stability indices, are considered in case studies in order to make a profound evaluation of the proposed approach in solving the OPF problem. This chapter provides an insight into the valve point effect to calculate the generating cost; a unified power flow controller (UPFC) is also added to improve the OPF stability. The complexity of OPF is increased due to valve point effect and hence efficient intelligent algorithm is required to solve that problem. FPA is found to be more efficient approach for solving this complex problem.

6.2   IMPLEMENTATION OF FPA FOR SOLVING OPF

Nearly a quarter of a million types of flowering plants are found in nature, and almost 80% of all plant species are flowering species. Flowering plants or angiosperms reproduce through pollination – a process of fertilization that involves transfer of pollen. The transfer of pollen is often linked with pollinators such as birds, insects, bats, and other animals. The two major forms of pollination are abiotic and biotic. Nearly 90% of flowering plants fit to biotic pollination (pollen is transferred by a pollinator such as insects and animals). The remaining 10% of pollination adapts abiotic form, which does not require any pollinators. Statistically, around 200,000 varieties of pollinators are available in the form of insects, birds, and bats. The best example of pollinator is honeybees. The honeybees can develop the so-called flower constancy. These pollinators have a tendency to visit particular flower species while avoiding other flower species. Such flower constancy with evolutionary benefits maximizes the transfer of flower pollen to the same or specific plants, thereby increasing the reproduction of the same flower species.

The concept of flower constancy is found to be advantageous for pollinators as well, since they can be sure that nectar supply is available with their limited memory and minimum cost of learning or exploring. Flower constancy may require less investment cost and more likely guaranteed intake of nectar. Pollination is of two types: self-pollination and cross-pollination. Self-pollination or local pollination occurs when the pollen from a flower fertilizes the same plant, and cross-pollination occurs when the pollen from one flower fertilizes another plant. Cross-pollination or global pollination requires either a biotic (bees and insects) or abiotic (wind, water, etc.) carrier to carry the pollen. Biotic cross-pollination may occur at long distance; since pollinators such as flies, bees, birds, and bats can fly a long distance, they are considered global pollinators. Furthermore, bees and birds may present Lévy flight behavior with jump or fly distance steps that obey a Lévy distribution. In addition, flower constancy can be used as an increment step by using the similarity or difference between two flowers. Figure 6.1 shows the pollination types in pictorial form.

Certain characteristics of the method of biological flower pollination were emulated while developing the FPA algorithm. These are summarized as follows:

- The global pollination occurs due to biotic transport of pollen, and the pollens are transported through Lévy flight phenomenon.
- Local pollination occurs as a result of abiotic transport of pollen.
- Flower constancy, also known as the reproduction probability, is proportional to the resemblance between two flowers.
- Both global and local pollination are controlled by the switching probability $p \in [0, 1]$. The value of p during local pollination can be significant due to the presence of abiotic pollinators such as wind.

In order to simplify the FPA algorithm, it was assumed that each flowering plant had only one flower that would emit a single pollen or gamete. The flower constancy or the reproduction probability during global pollination is calculated by using Equation (6.1):

$$x_i^{t+1} = x_i^t + F(x_i^t + g_{best}) \tag{6.1}$$

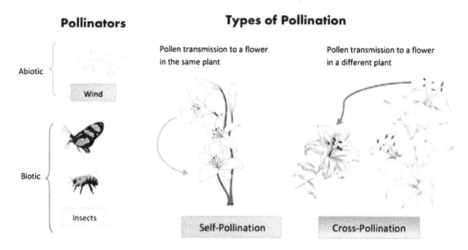

FIGURE 6.1 Types of pollination.

where x_i^t is the pollen i at iteration t, and g_{best} is the next best solution. F is the Lévy flight, which best describes the movement of insects. The length of steps during an insect locomotion is considered nonuniform, and it is described using the Lévy flight phenomenon.

Therefore, if $F > 0$, the Lévy distribution can be represented using Equation (6.2):

$$F \sim \frac{\lambda \Gamma(\lambda) \sin(\frac{\pi\lambda}{2})}{\pi} \frac{1}{s^{1+\lambda}}, \; s \gg s_0 > 0 \tag{6.2}$$

where $\Gamma(\lambda)$ represents the gamma function, and it is valid for larger steps.

On the other hand, local pollination is defined using Equation (6.3):

$$x_i^{t+1} = x_i^t + \phi(x_j^t - x_k^t) \tag{6.3}$$

where, x_j^t and x_k^t are pollens that emerged from two different flowers belonging to the same species. A generic flowchart summarizing the steps involved in developing FPA is illustrated in Figure 6.2.

6.3 CASE STUDY

The performance of the FPA for solving the OPF problems is clearly explained in this section. To substantiate the accurate performance of the proposed algorithm in optimizing the most complicated optimization problems, it is examined by case studies on IEEE 30 bus system. It is worth mentioning that the proposed approach has been implemented in MATLAB 2014A environment.

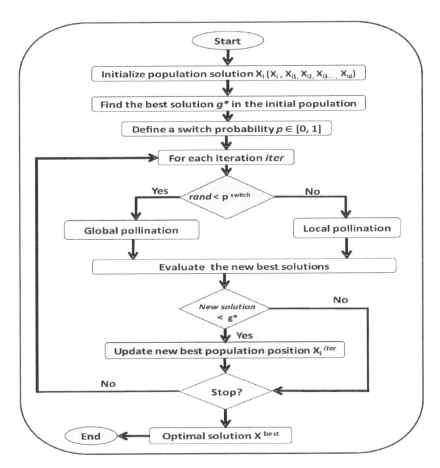

FIGURE 6.2 Flowchart summarizing the flower pollination algorithm.

CASE I GENERATING COST MINIMIZATION

The focus of Case 1 is to reduce the generating cost of the IEEE 30 bus system. For this case, the quadratic cost function with valve point loading effect is considered. Table 6.1 gives the comparative results between ABC and FPA approach for the test case with and without UPFC. The results from the table reflect that FPA performs better than ABC algorithm. The comparison between different algorithms and generating cost is shown in Figure 6.3. Figure 6.3 shows that FPA with UPFC gives minimum generating cost of $891.728/hour as compared with other algorithms.

CASE II EMISSION MINIMIZATION

The focus of Case 2 is to reduce the emission, which is a major threat to the entire world (global warming). The emission is due to burning of coal for power generation, which emits CO and CO_2. Therefore, emission should be reduced as far as possible. The FPA is executed for the test case with UPFC and without UPFC and the emission results are indicated in Table 6.2 along with ABC algorithm. From Table 6.2, it is

TABLE 6.1

Cost Minimization Objective for IEEE 30 Bus

Generation	Without UPFC		With UPFC
	ABC	FPA	FPA
P_{g1}	194.844	94.927	177.596
P_{g2}	51.992	62.791	74.902
P_{g3}	15.000	35.125	28.026
P_{g4}	10.000	43.351	12.070
P_{g5}	10.000	25.073	19.289
P_{g6}	15.657	27.542	31.687
UPFC-connected B/W buses	–	–	10–21
UPFC shunt power	–	–	$0 + j180.782$
UPFC series power	–	–	$6.022 + j1.282$
Cost ($/hour)	945.4495	934.3160	891.728

deduced that FPA with UPFC provides minimum emission in comparison with other algorithms. Figure 6.4 clearly shows the comparison between different algorithms and emission. In Figure 6.4, the emission reduced by UPFC is from 0.3752 to 0.2573 tons.

CASE III TRANSMISSION LOSS MINIMIZATION

The focus of case 3 is to reduce the transmission loss, which yet is another crucial parameter in OPF as due to reduction in loss, there will be considerable increase in

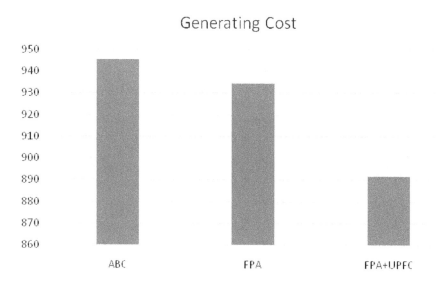

FIGURE 6.3   Comparison of different algorithms versus generating cost.

TABLE 6.2
Emission Minimization Objective for IEEE 30 Bus

Generation	Without UPFC		With UPFC
	ABC	FPA	FPA
P_{g1}	194.844	181.329	177.596
P_{g2}	51.992	39.043	74.902
P_{g3}	15.000	19.654	28.026
P_{g4}	10.000	17.380	12.070
P_{g5}	10.000	15.424	19.289
P_{g6}	15.657	21.666	31.687
UPFC-connected B/W buses	–	–	10–21
UPFC shunt power	–	–	$0 + j180.782$
UPFC series power	–	–	$6.022 + j1.282$
Emission (ton/hour)	0.42377	0.3752	0.2573

generation. The FPA is implemented for the test case and the results are tabulated in Table 6.3. Based on Table 6.3, the FPA approach provides better loss minimization in contrast with ABC approach (Figure 6.5).

CASE IV Voltage Stability Index Minimization

Case 4, namely, voltage stability index, is an important issue to be addressed with respect to power system stability. For effective supply of electric power, the voltage

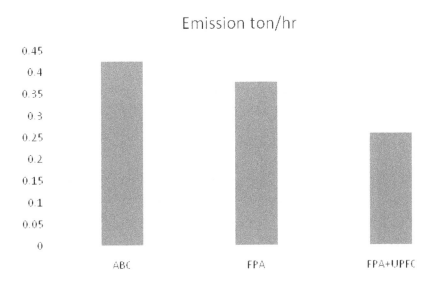

FIGURE 6.4 Comparison of different algorithms versus emission.

TABLE 6.3

Loss Minimization Objective for IEEE 30 Bus

Generation	Without UPFC		With UPFC
	ABC Algorithm	FPA	FPA
V_{g1}	1.0227	1.009	0.981
V_{g2}	1.0035	1.001	0.981
V_{g3}	1.0252	0.973	0.938
V_{g4}	1.0076	0.976	0.968
V_{g5}	0.9821	1.020	0.944
V_{g6}	1.1000	0.977	0.975
T_1	1.0500	0.926	1.096
T_2	1.1000	0.993	1.047
T_3	0.9625	1.023	1.085
T_4	0.9000	1.079	0.905
UPFC-connected B/W buses	–	–	10–17
UPFC shunt power	–	–	$0 + j50.2579$
UPFC series power	–	–	$11.91 + j6.01$
Loss (MW)	14.0928	11.0975	9.2670

has to be maintained within the tolerance. The voltage stability index is measured in terms of L-index and the values are determined using the FPA; these are tabulated in Table 6.4. From Table 6.4, it is inferred that FPA with UPFC provides optimized result in comparison with FPA without UPFC and ABC approach.

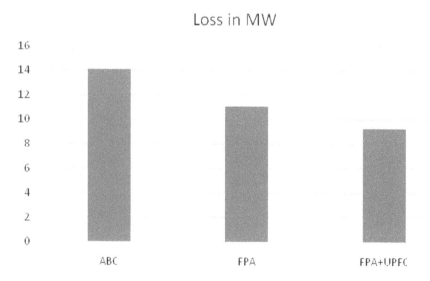

FIGURE 6.5   Comparison of different algorithms versus transmission loss.

TABLE 6.4

Voltage Stability Index Objective for IEEE 30 Bus

Particulars	Without UPFC		With UPFC
			FPA
	ABC	FPA	FPA-DE
V_{g1}	1.0227	1.009	1.024
V_{g2}	1.0035	1.001	1.010
V_{g3}	1.0252	0.973	1.002
V_{g4}	1.0076	0.976	1.014
V_{g5}	0.9821	1.020	1.008
V_{g6}	1.1000	0.977	1.016
UPFC-connected B/W buses	–	–	3–4
UPFC shunt power	–	–	0+j130.878
UPFC series power	–	–	0+j4.4473
L-index	0.1431	0.1320	0.1146

In Figure 6.6, the UPFC reduces the L-index value from 0.1320 to 0.1146. From the stability viewpoint, the L-index should be as low as possible. The impact of minimization of VSI leads to increase in stability of the power system.

Furthermore, Table 6.1 tabulates a comparison between the best costs obtained by FPA and the existing approaches in the literature. It is clear that the FPA with UPFC approach provides less fuel cost (shown graphically in Figure 6. 3) and also obtain a

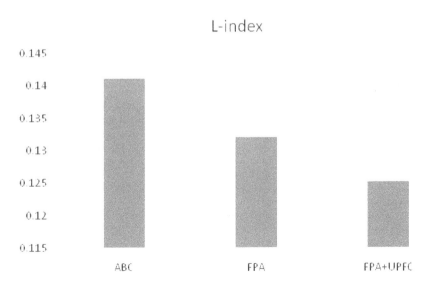

FIGURE 6.6   Comparison of different algorithms versus voltage stability index.

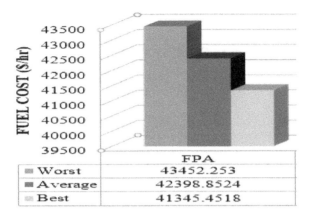

FIGURE 6.7   Fuel cost optimization for 50 trails.

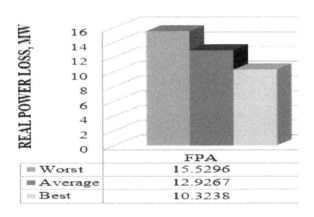

FIGURE 6.8   Real power loss optimization for 50 trails.

better optimal solution compared to the other existing approaches, which proves its aptitude in handling constraints in the OPF problem successfully.

To take care of different initial condition, the algorithm is executed for 50 autonomous runs and the best, mediocre, and worst solutions are given in the graphical representation, as shown in Figures 6.7 and 6.8.

6.4   CONCLUSION

The novel nature-inspired FPA approach with FACTS devices is implemented for OPF. The results obtained using FPA is a clear evidence for the potential use of FPA. Performance of this algorithm is evaluated using IEEE 30 bus. These computed results using FPA are compared with algorithms. With respect to the accuracy, robustness, speed of convergence, and complexity, the FPA seems to have very good performance in comparison with algorithms. FPA with FACTS devices gives minimum generating cost, emission, and loss and improves stability index for the objective function with and without valve point loading.

6.5 SUMMARY

i. FPA and their application are discussed in detail and the flowchart is drawn for solving OPF problem.
ii. Conducted case study and the objective functions are optimized using FPA approach.
iii. The execution time is less with good accuracy and the results show that FPA is better than other approaches.

REFERENCES

Abdelaziz, AY, Ali, ES & Abd Elazim, SM, 2016, 'Optimal sizing and locations of capacitors in radial distribution systems via flower pollination optimization algorithm and power loss index', Engineering Science and Technology, an International Journal, vol. 19, no. 1, pp. 610–618. https://www.sciencedirect.com/science/journal/22150986/19/1

Blum, C & Roli, A, 2003, 'Metaheuristics in combinatorial optimization: Overview and conceptual comparison', ACM Computing Surveys, vol. 35, no. 3, pp. 268–308.

Cuevas, E & Cienfuegos, M, 2014, 'A new algorithm inspired in the behavior of the social-spider for constrained optimization', Expert Systems with Applications, vol. 41, no. 2, pp. 412–425.

Deb S & Goswami AK, 2016, 'Congestion management by generator real power rescheduling using flower pollination algorithm'. In: Control, Instrumentation, Energy & Communication (CIEC), 2nd International Conference, IEEE, pp. 437–441.

Dubey, HM, Pandit, M & Panigrahi, BK, 2015, 'Hybrid flower pollination algorithm with time-varying fuzzy selection mechanism for wind integrated multi-objective dynamic economic dispatch', Renewable Energy, vol. 83, pp. 188–202. https://www.science-direct.com/science/journal/09601481 and https://www.sciencedirect.com/science/journal/09601481/83/supp/C

Hegazy O, Soliman OS & Salam MA, 2015, 'Comparative study between FPA, BA, MCS, ABC, and PSO algorithms in training and optimizing of LS-SVM for stock market prediction', International Journal of Advanced Computer Research, vol. 5, no. 18, pp. 35–45.

Karaboga, D, Gorkemli, B, Ozturk, C & Karaboga, N, 2014, 'A comprehensive survey: artificial bee colony (abc) algorithm and applications', Artificial Intelligence Review, vol. 42, no. 1, pp. 21–57.

Kazemian, M, Ramezani, Y, Lucas, C & Moshiri, B, 2007, 'Swarm clustering based on flowers pollination by artificial bees'. In: Swarm Intelligence in Data Mining. DOI: 10.1007/978-3-540-34956-3_8

Kumar, PRJ & Natarajan, R, 2017, 'A new global maximum power point tracking technique for solar photovoltaic (PV) system under partial shading conditions (PSC)', Energy, vol. 118, no. 1, pp. 512–525.

Ouadfel, S & Taleb-Ahmed, A, 2016, 'Social spiders optimization and flower pollination algorithm for multilevel image thresholding: A performance study', Expert Systems with Applications, vol. 55, pp. 566–584.

Rathasamuth, W & Nootyaskool, S, 2016, 'Comparison solving discrete space on flower pollination algorithm, PSO and GA', 2016 8th International Conference on Knowledge and Smart Technology (KST), DOI: 10.1109/KST.2016.7440499.

Sakib, N. Kabir, MWU, Rahman, MS & Alam, MS, 2014, 'Article: A comparative study of flower pollination algorithm and bat algorithm on continuous optimization problems', International Journal of Applied Information Systems, vol. 7, no. 9, pp. 19–20.

Walton, S, Hassan, O, Morgan, K & Brown, MR, 2011, 'Modified cuckoo search: A new gradient free optimisation algorithm' Chaos, Solitons & Fractals, vol. 44, no. 9, pp. 710–718.

Yang, X-S, 2012, 'Flower pollination algorithm for global optimization', LNCS, vol. 7445, pp. 240–249.

7 Real-Coded Genetic Algorithm Differential Evolution (RGA-DE) Based OPF

LEARNING OUTCOME

i. To provide a detailed study of real-coded genetic algorithm differential evolution (RGA-DE) approach.
ii. To study the RGA-DE algorithm that includes flexible alternating current transmission system (FACTS) devices: static synchronous compensator, static synchronous series compensator, unified power flow controller for optimization.
iii. RGA involves selection and crossover, whereas DE involves mutation operator.
iv. To compare the fuel cost and real power loss by RGA approach with different FACTS devices.

7.1 INTRODUCTION

The optimal power flow (OPF) objective is to determine the minimum operating cost without violating the limits and to be operated in secure state. The traditional mathematical programming approach available to solve the OPF is linear programming (LP), quadratic programming (QP), and nonlinear programming (NLP) methods. However, the classical approaches are not superior in finding the global optimum and are also highly sensitive to initial condition. But the metaheuristic algorithms overcome this local minimum values and identify global optimal values. In the past few years, evolutionary-based algorithms have been developed to rule these restrictions out to find approximate solutions to the different kinds of optimization problems. Considering the recent literature, various heuristic and metaheuristic algorithms have been implemented to solve the OPF. Some of the algorithms that are employed are artificial bee colony (QCABC) (Karaboga and Gorkemli 2019); krill herd algorithm (KHA) (Gandomi and Alavi 2012); symbiotic organisms search (SOS) (Cheng and Prayogo 2014); imperialist competitive algorithm (ICA) (Atashpaz-Gargari and Lucas 2007); fuzzy harmony search algorithm (FHSA) (Peraza et al. 2016); and DE (Suganthan 2012). Alsac and Stott (1974) extend conventional Dommel-Tinney approach to determine the solution for OPF problem and for that proposed they used IEEE 30 bus system. Based on their approach, IEEE 30 bus system is identified as standard test case for most of the OPF problem solutions.

(Bakirtzis et al. 2002) used enhanced genetic algorithm, which is a binary coded GA, for reducing enhanced number of bits to 25 bits. Genetic algorithm (GA) was introduced by John Holland in the 1970s. It mimics Darwin's evolution theory, and it uses the concept of survival of fitness. GA is widely used for solving nonlinear, nonconvex optimization problem. GA is divided into binary coded GA and real-coded GA (RGA). In RGA, all variables are real numbers as in real-world application. The objective function in RGA is taken as fitness function, control variables are taken as genes, and a group of control variables is considered as chromosome. The minimum and maximum limits for control and dependent variables form the boundary of the solution space. RGA explores this space and finds the global best value. Like nature, it optimizes the chromosome fitness through generation by generation RGA and tunes the genes to the best optimal value. A set of chromosomes is termed as population. RGA evaluates each chromosome in the population to keep better chromosomes and replaces unfit chromosomes in each generation, and finally it brings the best chromosome that has global optimum value.

The three main operation of GA are selection, crossover, and mutation.

a. *Selection:* Selection is the process of inspecting high-fitness chromosome in the current population and provides opportunity in succeeding generation. The various types of selection schemes are tournament selection, truncation selection, linear ranking selection, elitist selection, and roulette wheel selection.

b. *Crossover:* Crossover is the process of interchanging information (genes) among the parents in the mating pool. It can be classified as single-point, tow-point, and multipoint crossover.

c. *Mutation:* Mutation is the process of toggle in the specified bit of a gene in the chromosome for binary coded GA. In RGA, particular gene value is changed without violating its limits. Mutation is required for divergent search in the solution space. The probability of mutation in RGA is less as compared to crossover.

d. *Stopping criteria:* The evolution of generation is stopped by either maximum number of generation or solution convergence. If there is an improvement of solution in every generation, then convergence criterion may implement; or else if negligible improvement is seen, then maximum generation might be the best choice.

The concept of DE was proposed by Storn and Price in 1994 [Bhattacharya and Chattopadhyay 2010].

The capability of DE is to optimize nonlinear, noncontinuous, and nondifferential real-world problems. In comparison with other population-based heuristic algorithms, DE emphasizes on mutation than crossover. Real-valued control variables are grouped that form a vector, and a group of vectors is called population.

For mutation, it uses randomly selected vectors in the same population. This mutation helps the vector to move toward the global optimum [Elango and Senthilkumar 2018; Hariharan and Kuppusamy 2016; Hariharan and Sundaram 2015]. The processes of mutation and crossover produce new vectors, and the selection process selects the best vectors based on the selection criterion.

a. *Mutation:* The objective of mutation is to enable search diversity in the solution space and to direct the vector to approach global optimal solution. Commonly used four types of mutation are DE/rand/1, DE/rand/2, DE/best/1, and DE/best/2. For this mutation process, two to five vectors apart from target vector are chosen; for last two types of mutation rule, best vector in the current population is used.

b. *Crossover:* Crossover aims at reinforcing prior successes to current population. Binomial or exponential technique is used for this crossover. In binomial approach, a crossover constant is used to determine the importance of control variable in mutated or target vector. The outcome of the crossover process is trail vector.

c. *Selection:* Selection is the process of selecting either a target vector or a trail vector based on their objective value.

d. *Stopping criteria:* The maximum number of iteration is the commonly used stopping criterion. Convergence criterion may be used if the last few iterations do not have any improvement.

Hybrid RGA-DE

Hybrid RGA-DE algorithm utilizes the advantages of GA and DE to determine the global optimal value. The selection and crossover are chosen from GA and mutation is chosen from DE; maximum generation is considered as stopping criterion.

This hybridization of algorithms provides better solution than the individual algorithms.

a. *Initialization:* The first step is the selection of random initial values for all control variables within its minimum and maximum limits.

b. *Selection:* This step involves the process of choosing high-fitness chromosome from the mating pool. Here versatile roulette wheel selection is used.

c. *Crossover:* It is the process of sharing important information among the fittest chromosomes in the mating pool. Single-point crossover is utilized in this hybrid algorithm.

d. *Mutation:* To improve mutation process, "DE/rand/1" mutation rule is used. Mutation is carried out for all chromosomes in the population. A selected chromosome for mutation is called target vector and apart from target vector 3, more distinct chromosomes are selected to perform mutation.

e. *Stopping criteria:* The execution of selection, crossover, and mutation operation for a population is known as one generation. This heuristic algorithm is an iterative algorithm in which the process is repeated again and again till the stopping criteria. In hybrid algorithms, these iterations are called generations and the maximum number of generation is used as stopping criterion.

7.2 HYBRID RGA-DE ALGORITHM BASED OPF

Control variables are encoded into genes, and a set of these genes is called chromosome. Initial population of chromosome is generated from the solution space and then selection, crossover, and mutation operators are used to find the next-generation population. For each chromosome, an objective value is calculated, which is known as fitness of the chromosome. Solution space is subjected to constraints, namely, equality and inequality constraints. Power balance equation becomes equality constraint, and limits on control and dependent variables form inequality constraints.

To optimize OPF problem, the control variables, real power generation, generator bus voltages, and transformer tap position are considered. The limits on these control variables form prime constraints in addition to power balance condition. To use actual values of these control variables, RGA is used. The process of implementing hybrid RGA-DE algorithm involves encoding, define fitness function, selection, crossover, mutation, and stopping criteria.

7.2.1 Encoding

Encoding is the process of converting objective function into fitness function and the decision variable into gene. Actual values of control variables are taken from chromosome as like RGA. A set of genes is a chromosome and a set of chromosomes is called population. Population size depends on problem complexity. Initial population is created by generating random values of genes within the solution space. The population size is fixed and does not change till the solution is reached.

7.2.2 Fitness Function

Generation cost function is the objective function of OPF. Fitness function of hybrid RGA-DE includes all the objective functions and penalty functions if any. It evaluates fitness function for each chromosome in the population. Based on the evaluation, either the chromosome is retained or replaced by another chromosome. Chromosomes having high fitness replace low-fitness chromosomes.

7.2.3 Selection

Selection implements the natural evaluation of survival of the fittest according to the fitness function. It provides more possibility for reproduction to the high-fitness chromosomes and tries to remove the low-fitness chromosomes. RGA selection process is used in hybrid RGA-DE algorithm. The different types of selection schemes in RGA are as follows:

- Tournament selection
- Truncation selection
- Linear ranking selection
- Exponential ranking selection
- Elitist selection
- Roulette wheel selection

In this research work, hybrid RGA-DE uses roulette wheel selection to select the fitness chromosome from the mating pool, which participates in crossover operation.

7.2.4 CROSSOVER

Crossover is the process of interchanging information (genes) among the parent chromosomes in the mating pool. The crossover can be classified as single-point, two-point, and multipoint crossover. In the single-point crossover, after specified point the genes of the two parents are interchanged. In the two-point crossover, the genes in between two specified points are interchanged.

7.2.5 MUTATION

Hybrid RGA-DE algorithm uses DE algorithm mutation process. The objective of mutation is to enable search diversity in the solution space as well as to direct the existing vectors with suitable amount of parameter variation in a way that will lead to better results at a suitable time. It keeps the search robust and explores new areas in the search domain. Target vector is selected based on fitness function to find mutated vector by using randomly selected vector from the population other than target vector. Four types of commonly used mutation are given in Equations (7.1)–(7.4):

$$\text{DE / rand / 1 / bin}: X_{r1mutated} = X_{r1} + SF * (X_{r2} - X_{r3}) \tag{7.1}$$

$$\text{DE / rand / 2 / bin}: X_{r1mutated} = X_{r1} + SF * (X_{r2} - X_{r3}) + SF * (X_{r4} - X_{r5}) \tag{7.2}$$

$$\text{DE / best / 1 / bin}: X_{r1mutated} = X_{best} + SF * (X_{r1} - X_{r2}) \tag{7.3}$$

$$\text{DE / best / 2 / bin}: X_{r1mutated} = X_{best} + SF * (X_{r1} - X_{r2}) + SF * (X_{r3} - X_{r4}) \tag{7.4}$$

where
   X_{r1} is the target vector
   $X_{r1mutated}$ is the mutated vector
   X_{best} is the best optimal solution in the population
   SF is the scaling actor
   $r1$ to $r5$ are random vector positions in population
   $r1 \neq r2 \neq r3 \neq r4 \neq r5$

The first two mutation rules given in Equations (7.1) and (7.2) are called random vector mutation rule, and the other two mutation rules given in Equations (7.3) and (7.4) are called best vector based mutation rule. Appropriate scaling factor should be decided based on problem domain and its range is from 0 to 1; in this work it is taken as 0.7. High value of scaling factor may decrease in convergence speed but escapes from local minima. Equation (7.1) is used to generate mutated vector for target vector by using target vector, scaling factor, and other two randomly selected vectors from the population. To induce more diversity, four more random vectors are used, as given in Equation (7.2). In these two equations, target vector and other randomly selected vectors are used. To reinforce best vector in the population, Equations (7.3) and (7.4) are used.

7.2.6 Stopping Criteria

Hybrid RGA-DE improves problems' solution iteration by iteration and the iteration has to be stopped when either the problem is converged or the iteration reaches its maximum value. Stopping of iteration is important to provide solution for time complexity.

7.3 IMPLEMENTATION OF RGA-DE ALGORITHM FOR SOLVING OPF

The procedure for hybrid RGA-DE to solve OPF is given below:

1. Select control variables of OPF, these are selected as genes of a chromosome.
2. Create initial population.
3. Find fitness of chromosomes.
4. Select chromosome using roulette wheel selection for mating.
5. Perform crossover operation as in RGA.
6. Perform mutation operation as in DE.
7. Generate new population for next generation.
8. Repeat steps 4–7 until stopping criterion is satisfied.
9. Print the optimal result after stopping criterion is satisfied.

7.4 FLOWCHART

The flowchart for solving OPF using hybrid RGA-DE is shown in Figure 7.1.

7.5 CASE STUDY AND RESULTS

For validating the performance of the hybrid RGA-DE algorithm, IEEE 30 bus system is considered. For all cases, 100% loading condition is considered. For this system, the control variables are as follows: five real power generation, six generator bus voltages, four transformer tap positions, two control variables for the flexible alternating current transmission system (FACTS) devices STATCOM and SSSC, and four control variables for UPFC. For STATCOM or SSSC, one control variable is position and another one is the size of the device.

In case of UPFC, two control variables for the line that is nowhere connected and two control variables for shunt and series converter size. Hence, 17 control variables for STATCOM and SSSC and 19 control variables for UPFC are used. These control variables are termed as genes used to form a chromosome. Twenty such chromosomes form the population. This population evolves iteration by iteration to find global optimal solution. The maximum number of iteration it required to evolve is taken as 30 iterations. Crossover and mutation constants are taken as 0.7 and 0.01, respectively. The results of RGA-DE for three FACTS devices are given in Table 7.1. Table 7.2 presents the percentage of reduced loss and fuel cost using STATCOM, SSSC, and UPFC, respectively.

TABLE 7.1
Comparison of FACTS Devices in Hybrid RGA-DE Algorithm

Control Variables	RGA-DE with STATCOM	RGA-DE with SSSC	RGA-DE with UPFC
P_{G1} (MW)	188.29	186.91	178.12
P_{G2} (MW)	36.183	34.53	33.723
P_{G3} (MW)	31.752	29.41	28.652
P_{G4} (MW)	12.536	12.812	16.934
P_{G5} (MW)	14.103	14.18	15.874
P_{G6} (MW)	13.696	13.124	14.542
V_{G1} (pu)	1.02	1.03	1.01
V_{G2} (pu)	1.02	1.01	1.00
V_{G3} (pu)	1.01	0.99	1.05
V_{G4} (pu)	1.01	0.98	1.00
V_{G5} (pu)	1.00	1.01	0.97
V_{G6} (pu)	1.01	1.00	1.01
T_1 (pu)	0.997	0.952	0.967
T_2 (pu)	0.959	0.988	0.995
T_3 (pu)	1.02	1.04	0.992
T_4 (pu)	0.991	0.921	0.993
Injected shunt power (MVAr)	48	–	85
Injected series power (MW)	–	5.239	6.375
Position	2	6	2-1
Total power demand, MW	294.42	294.061	293.705
Loss (MW)	11.02	10.661	10.305
Cost ($/hour)	817.388	798.441	792.755

Figure 7.2 gives the real power loss convergence curve of OPF with STATCOM device and the loss converged on 11.02 MW at 96th iteration. The convergence curve for the STATCOM with the objective of minimizing generating cost is given in Figure 7.3. Here the generating cost converged on $817.388/hour at 94th iteration. Time taken to get convergence result is 34 seconds.

Figure 7.4 gives the real power loss convergence curve of OPF with SSSC device and the loss converged on 10.661 MW at 92nd iteration. The convergence curve for the SSSC with the objective of minimizing generating cost is given in Figure 7.5. Here the generating cost converged on $798.441/hour at 87th iteration. Time taken to get convergence result is 30 seconds.

Figure 7.6 shows the real power loss convergence curve of OPF with UPFC device and the loss converged on 10.305 MW at 72nd iteration. The convergence curve for the UPFC with the objective of minimizing generating cost is given in Figure 7.7. Here the generating cost converged on $792.755/hour at 81st iteration. Figure 7.8 gives comparison of the total power generation of all generators with FACTS devices. Figures 7.9 and 7.10 give the total real power loss and fuel cost comparison with FACTS devices, respectively. Time taken for convergence result is 32 seconds.

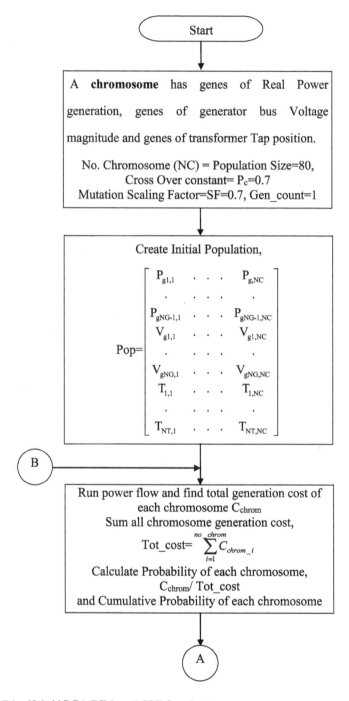

FIGURE 7.1   Hybrid RGA-DE-based OPF flowchart.

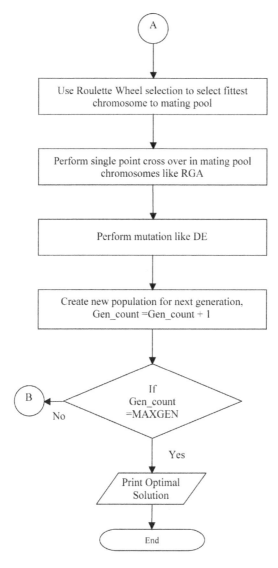

FIGURE 7.1   (*Continued*)

TABLE 7.2

Comparison of FACTS Devices with Reduced Loss Percentage and Fuel Cost Percentage

FACTS Devices	Reduced Loss Percentage (%)	Reduced Fuel Cost Percentage (%)
SSSC	3.25	2.31
UPFC	6.48	3.01

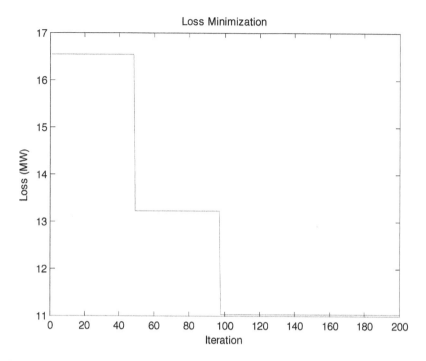

FIGURE 7.2   Real power loss convergence curve of OPF with STATCOM.

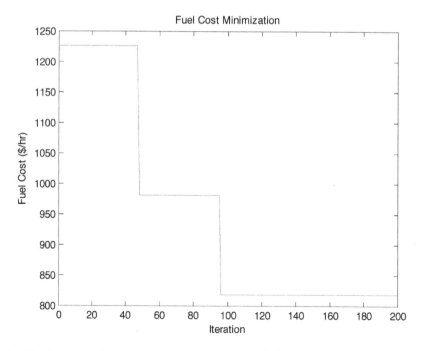

FIGURE 7.3   Generation cost convergence curve of OPF with STATCOM.

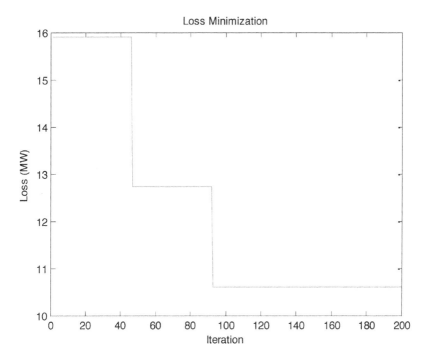

FIGURE 7.4   Real power loss convergence curve of OPF with SSSC.

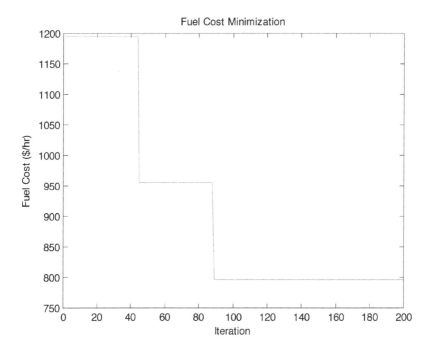

FIGURE 7.5   Generation cost convergence curve of OPF with SSSC.

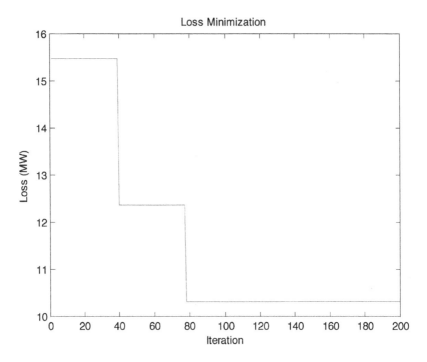

FIGURE 7.6   Real power loss convergence curve of OPF with UPFC.

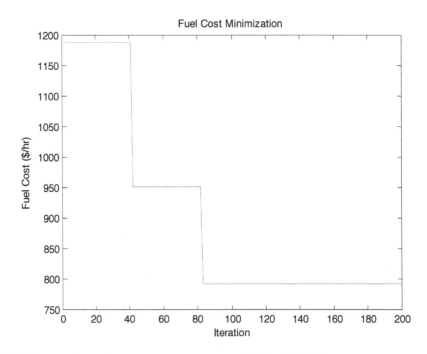

FIGURE 7.7   Generation cost convergence curve of OPF with UPFC.

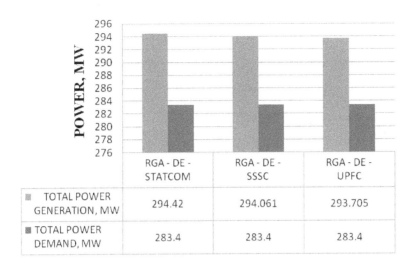

	RGA - DE - STATCOM	RGA - DE - SSSC	RGA - DE - UPFC
▓ TOTAL POWER GENERATION, MW	294.42	294.061	293.705
▓ TOTAL POWER DEMAND, MW	283.4	283.4	283.4

FIGURE 7.8 Total real power generation of all generators with FACTS devices.

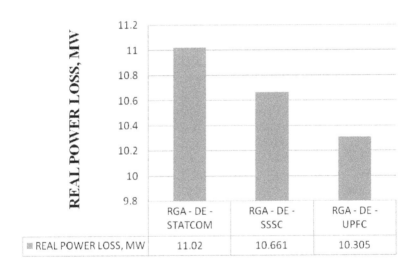

	RGA - DE - STATCOM	RGA - DE - SSSC	RGA - DE - UPFC
▓ REAL POWER LOSS, MW	11.02	10.661	10.305

FIGURE 7.9 Total real power loss comparisons with FACTS devices.

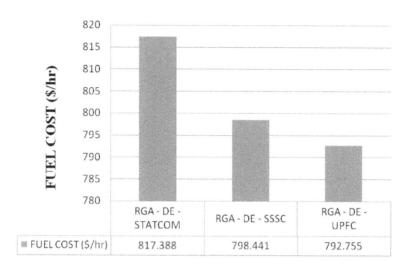

■ FUEL COST ($/hr)	RGA - DE - STATCOM	RGA - DE - SSSC	RGA - DE - UPFC
FUEL COST ($/hr)	817.388	798.441	792.755

FIGURE 7.10   Fuel cost comparisons with FACTS devices.

7.6   CONCLUSION

In this chapter, the hybrid RGA-DE approach is applied to solve OPF problem. It has three main operators: selection, crossover, and mutation. Selection and crossover are taken from RGA algorithm and mutation is taken from DE to develop this hybrid RGA-DE. Since hybrid RGA-DE algorithm has advantages of RGA and DE algorithm, it gives better optimal solution as compared to RGA and DE algorithms. To evaluate the performance of developed algorithm, IEEE 30 bus system is considered.

Based on 100% loading, condition is considered in RGA-DE algorithm to obtain optimal power generation, loss, and fuel cost. The percentage reduction of loss for SSSC is 3.25% and for UPFC it is 6.48%. Likewise, the reduction in the generating cost by SSSC is 2.31% and by UPFC it is 3.01%. Hybrid RGA-DE gives minimum generating cost for the objective function with and without valve point loading as compared to RGA and DE algorithms.

7.7   SUMMARY

  i. OPF with STATCOM analysis done by RGA-DE approach for minimization of generating cost and reduction of losses.
 ii. OPF with SSSC analysis done by RGA-DE approach for both minimization of generating cost and minimization of loss.
iii. OPF with UPFC analysis done by RGA-DE approach for minimization of generating cost and reduction of loss.
 iv. A comparative analysis has been done between three FACTS devices, and UPFC provides minimum generating cost of $591.11/hour.

REFERENCES

Alsac, O & Stott, B, 1974, 'Optimal load flow with steady-state security', IEEE Transactions on Power Apparatus and Systems, vol. PAS-93, no. 3, pp. 745–75.

Atashpaz-Gargari, E & Lucas, C, 2007, 'Imperialist competitive algorithm: An algorithm for optimization inspired by imperialistic competition', IEEE Congress on Evolutionary Computationm, IEEE, 2007.

Bakirtzis, AG, Biskas, PN, Zoumas, CE & Petridis, V, 2002, 'Optimal power flow by enhanced genetic algorithm', IEEE Transactions on Power Systems, vol. 17, no. 2, pp. 229–236.

Cheng, M-Y & Prayogo, D, 2014, 'Symbiotic organisms search: A new metaheuristic optimization algorithm', Computers & Structures, vol. 139, pp. 98–112.

Gandomi, AH & Alavi, AH, 2012, 'Krill Herd: A new bio-inspired optimization algorithm', Communications in Nonlinear Science and Numerical Simulation, vol. 17, no. 12, pp. 4831–4845.

Karaboga, D & Gorkemli, B, 2019, 'Solving traveling salesman problem by using combinatorial artificial bee colony algorithms', International Journal on Artificial Intelligence Tools, vol. 28, no. 1.

Peraza, C, et al., 2016, 'A new fuzzy harmony search algorithm using fuzzy logic for dynamic parameter adaptation', Algorithms, vol. 9, no. 4, p. 69.

Suganthan, PN, 2012, 'Differential evolution algorithm: Recent advances'. In: Dediu, AH, Martín-Vide, C, Truthe, B (eds), Theory and Practice of Natural Computing, TPNC 2012, Lecture Notes in Computer Science, vol. 7505, Springer.

8 Hybrid FPA-DE-Based OPF

LEARNING OUTCOME

i. A clear understanding of hybrid flower pollination algorithm differential evolution (FPA-DE) to solve various optimal power flow (OPF) problems.
ii. To study the performance of hybrid FPA-DE approach using IEEE 30 and 57 bus systems.
iii. To calculate the optimized value of fuel cost, loss minimization, emission, and voltage stability index.
iv. Numerical results of hybrid algorithms compared with other algorithms.

8.1 INTRODUCTION

The optimal power flow (OPF) is noticeable and integral tool in power systems operation and control. There are few evolutionary-based algorithms developed and applied to solve OPF. The algorithms such as particle swarm optimization (PSO) (Abido and Al-Ali 2009); artificial bee colony (ABC) (Bhattacharya and Chattopadhyay 2010); genetic algorithm (GA) (Chung 2010), differential evolution (DE) (Li et al. 2010); imperialist competitive algorithm (ICA) (Hariharan and Mohana Sundaram 2016); fuzzy harmony search algorithm (FHSA) (Yan and Li 2010); backtracking search algorithm (BSA) (Herbadji et al. 2016); firefly algorithm (FFA) cuckoo optimization algorithm (COA) (Ponnin Thilagar and Harikrishnan 2015); moth swarm algorithm (MSA), flower pollination algorithm (FPA) (Prathiba et al. 2014); and improved colliding bodies optimization (Sahu et al. 2015) are used to solve the OPF as single-objective approach. The single-objective approach can produce one optimal solution in a single run.

The multiobjective optimization technique is to optimize a set of objective functions simultaneously to address difficult problems. However, the objective functions are not in line with each other, so for solving multiobjective function, there will be a set of optimal solutions rather than one. These multiobjective optimization algorithms are proven tools in order to explore the search space entirely while solving the multiobjective optimal power flow (MOOPF) problems.

In the past literature survey, few algorithms such as gravitational search algorithm (GSA), ABC, and modified teaching-learning based optimization (MTLBO-DE) have solved OPF as MOOPF considering total fuel cost, power transmission losses, and voltage stability enhancement. The multiobjective optimization OPF problem is a highly complex and nonlinear problem with noncontinuous and nondifferentiable objective functions. Hence, it is important to adapt accurate algorithm for solving MOOPF. The formula of hybridization algorithm is adopted and in this chapter the insight of hybrid FPA-DE algorithm is clearly explained.

FPA has property of selection and crossover operation without mutation. The purpose of mutation is to provide global optimal solution and hence DE characterized with good mutation operator can be combined with FPA and used for solving MOOPF with valve point effect with unified power flow controller (UPFC). Another strategy to streamline the efficiency and performance of the SPSO algorithm is spreading the diversity of the population, which is crucially important to cover the search space utterly. To this end, the DE algorithm is hybridized with the modified SPSO algorithm in such a manner that they constitute a unify optimization algorithm to modify the diversification of the population, to increase the chance of escaping from local optima, and also to make a cogent equilibrium between exploration and exploitation stages.

Differential evolution algorithm is used to solve a number of engineering problems owing to its ease of implementation and robust performance. They are also used for solving conventional engineering design problems such as designing IIR filters (Sahu et al. 2015) and synthesizing six-bar linkages (Shilaja and Ravi 2016). The DE is an excellent technique for determining the optima using randomly selected vectors, but it lacks the guidance to move toward the global optimum. On the other hand, the FPA provides global and local search strategies, but does not perform as well as the DE. Considering the above facts, the combination of differential evolution algorithm with some modifications in the operator of FPA furnished with two adaptive dynamic weights enables local and global searching.

The various multiobjectives of OPF handled by the FPA-DE algorithm are reduction in generating cost, emission, active power line losses, and voltage stability index (VSI). FPA is inspired by flowering plants. Further, pollination can be classified as cross-pollination and self-pollination. Global pollination adopts the rule of biotic and cross-pollination. The global pollination is presented by Equation (8.1):

$$F_i^{n+1} = F_i^n + \gamma L(F_g - F_i^n) \tag{8.1}$$

where

F_i^n is the ith pollen at t iterations

γ is the scaling factor that controls the step size

L is the step size that corresponds to the strength of pollination, which is a Lévy weight based step size

F_g is the current best solution at current iteration

From the abiotic and self-pollination processes, the local pollination is derived. The flower constancy is equivalent to the reproduction probability. The local pollination is carried out by Equation (8.2) as follows:

$$F_i^{n+1} = F_i^n + \epsilon\left(F_j^n - F_k^n\right) \tag{8.2}$$

where

F_i^n is the ith pollen at t iterations

$\in$ takes a value of [0,1]

F_j^n and F_k^n are pollens from different flowers from the same plant.

The hybridization process involves FPA, and the mutation process is added, derived from the DE algorithm.

This combination of FPA and DE mutation constitutes hybrid algorithm and is applied for solving MOOPF problem. The purpose of mutation is to enable search diversity in the solution space and also to direct the existing vectors with appropriate amount of parameter variation in a way that yields good results at a particular time. The selection of target vector is based on fitness function to find mutated vector through randomly selected vectors from the population other than target vector. The commonly applied mutation rules are given in Equations (8.3–8.6):

$$\text{DE/rand/1/bin}: X_{r1mutated} = X_{r1} + SF*(X_{r2} - X_{r3}) \tag{8.3}$$

$$\text{DE/rand/2/bin}: X_{r1mutated} = X_{r1} + SF*(X_{r2} - X_{r3}) + SF*(X_{r4} - X_{r5}) \tag{8.4}$$

$$\text{DE/best/1/bin}: X_{r1mutated} = X_{best} + SF*(X_{r1} - X_{r2}) \tag{8.5}$$

$$\text{DE/best/2/bin}: X_{r1mutated} = X_{best} + SF*(X_{r1} - X_{r2}) + SF*(X_{r3} - X_{r4}) \tag{8.6}$$

where

X_{r1} is the target vector
$X_{r1mutated}$ is the mutated vector
X_{best} is the best optimal solution in the population
SF is the scaling actor
$r1$ to $r5$ are random vector positions in population
$r1 \neq r2 \neq r3 \neq r4 \neq r5$

The two mutation rules given in Equations (8.3) and (8.4) are termed as random vector mutation rule, and the other two rules Equations (8.5) and (8.6) are known as best vector mutation rule. The scaling factor is an important parameter and an appropriate value can be selected based on problem domain; the values are chosen from 0 to 1, and the value selected for this problem is 0.7. If the scaling factor is high, it will decrease the speed of convergence and thereby local minimum is escaped. To generate the mutated vector for target vector, the scaling factor and the other two randomly selected vectors from the population equation, that is, Equation (8.3), is adopted. Another four random vectors are used to induce more diversity, as denoted in Equation (8.4). For identifying the best vector in the population, Equations (8.5) and (8.6) are used.

8.2 IMPLEMENTATION OF HYBRID FPA-DE ALGORITHM

The following steps should be carried out to execute the algorithm:

1. Create initial population of flowers.
2. Evaluate the fitness function of each flower.
3. The flower with highest fitness turns global flower.
4. Generate a random number for each flower.
5. If the generated random number is less than switch probability, go to step 7.
6. Apply global pollination and go to step 8.
7. Apply local pollination.
8. Execute mutation operation as in differential evolution.
9. Repeat steps 2–8 until stopping criterion is satisfied.
10. End with optimal solution.

8.3 FLOWCHART

Figure 8.1 shows the flowchart for the hybrid algorithm.

8.4 CASE STUDY I: SOLVING THE PROBLEM ON IEEE 30 BUS SYSTEM

A comprehensive analysis of the proposed OPF problems on different test systems such as IEEE 30 is provided in this section (Appendix 1).

8.4.1 GENERATING COST MINIMIZATION

The first case is mainly focused on reducing the generating cost of the IEEE 30 bus system. The hybrid FPA-DE algorithm is implemented using MATLAB and generating cost is shown in Table 8.1.With respect to Table 8.1, the hybrid FPA-DE algorithm with UPFC provides reduced generating cost of \$876.081/hour. The convergence curve of this FPA-DE algorithm is displayed in Figure 8.2. Here the cost converged on \$876.081/hour at 41st iteration. The time taken to get converge is 47 seconds. Figure 8.3 displays the comparison of various algorithms versus generating cost.

8.4.2 EMISSION MINIMIZATION

The second important case to be considered is minimization of emission of greenhouse gases. The emission of CO and CO_2 is produced by burning of coal for electric power generation. Considering the standard test case, emission is calculated for different algorithms and shown in Table 8.2. According to the table, FPA-DE with UPFC shows minimum emission of 0.2231 ton/hour in comparison with other algorithms. Figure 8.4 shows the emission convergence curve of hybrid FPA-DE algorithm with UPFC. From the emission cost convergence curve, the emission cost is converged on 0.2231ton/hour at 53rd iteration. The time taken for convergence is

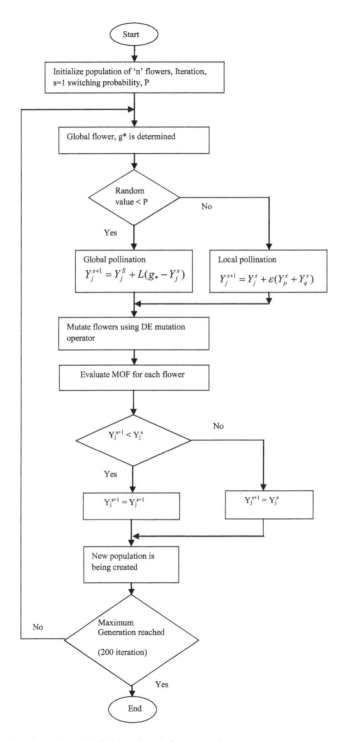

FIGURE 8.1 Flowchart of hybrid FPA-DE algorithm for OPF.

TABLE 8.1

Cost Minimization Objective for IEEE 30 Bus

	Without UPFC			With UPFC	
Generation	**ABC (Zhang and Li 2010)**	**FPA**	**FPA-DE**	**FPA**	**FPA-DE**
P_{g1}	194.844	94.927	108.624	177.596	118.0496
P_{g2}	51.992	62.791	69.350	74.902	46.870
P_{g3}	15.000	35.125	23.801	28.026	40.149
P_{g4}	10.000	43.351	28.211	12.070	39.193
P_{g5}	10.000	25.073	23.064	19.289	30.000
P_{g6}	15.657	27.542	37.229	31.687	17.757
UPFC-connected B/W buses	–	–	–	10–21	6–7
UPFC shunt power	–	–	–	$0+j180.782$	$0+j233.668$
UPFC series power	–	–	–	$6.022+j1.282$	$2.163+j12.525$
Cost ($/hour)	945.4495	934.3160	923.876	891.728	876.081

49 seconds. Figure 8.5 shows the comparison of various algorithms versus emission cost.

8.4.3 MINIMIZATION OF TRANSMISSION LOSS

Loss minimization is the third important parameter to be considered as loss reduction leads to increase in generation. Table 8.3 enlists the results obtained from FPA-DE algorithm with UPFC and a comparison is also made with other algorithms. From

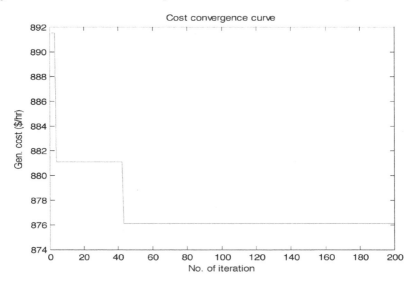

FIGURE 8.2   Generation cost convergence curve for FPA-DE with UPFC.

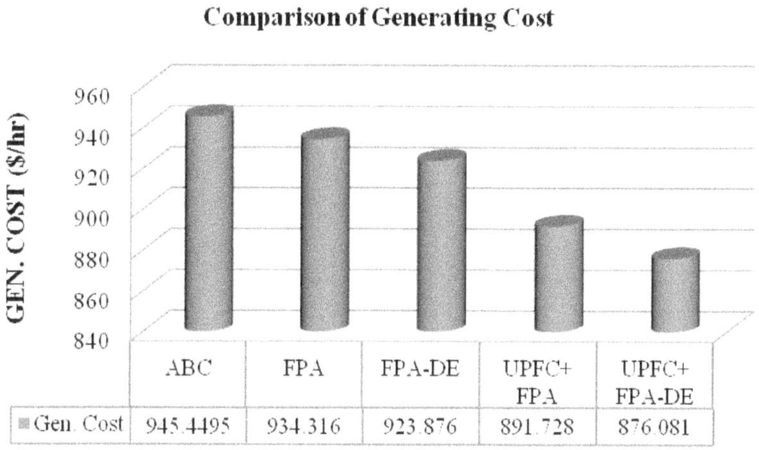

FIGURE 8.3 Comparison of various algorithms versus generating cost.

Table 8.3, it is evident that hybrid FPA-DE algorithm with UPFC provides less loss. The corresponding loss is converged on 5.4846 MW at eighth iteration, as shown in Figure 8.6. The time taken to achieve this convergence is 12 seconds. Figure 8.7 shows the comparison of various algorithms versus loss.

8.4.4 VSI MINIMIZATION

The fourth objective is the VSI, another major factor from stability point of view. VSI is measured in terms of L-index, which is given in Table 8.4. According to

TABLE 8.2

Emission Minimization Objective for IEEE 30 Bus

Generation	Without UPFC			With UPFC	
	ABC (Zhang and Li 2010)	FPA	FPA-DE	FPA	FPA-DE
P_{g1}	194.844	181.329	147.377	177.596	86.815
P_{g2}	51.992	39.043	61.460	74.902	80.000
P_{g3}	15.000	19.654	17.305	28.026	50.000
P_{g4}	10.000	17.380	35.625	12.070	53.071
P_{g5}	10.000	15.424	19.003	19.289	30.000
P_{g6}	15.657	21.666	12.498	31.687	36.082
UPFC-connected B/W buses	–	–	–	10–21	6–10
UPFC shunt power	–	–	–	$0 + j180.782$	$0 + j148.882$
UPFC series power	–	–	–	$6.022 + j1.282$	$59.797 + j4.459$
Emission (ton/hour)	0.42377	0.3752	0.3018	0.2573	0.2231

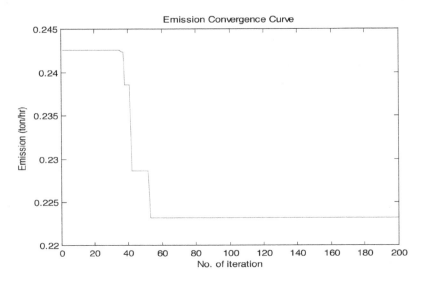

FIGURE 8.4   Emission convergence curve of hybrid FPA-DE algorithm with UPFC.

Table 8.4, FPA-DE approach with UPFC provides reduced voltage stability in comparison with another metaheuristic approach.

Figure 8.8 shows the voltage stability L-index convergence curve for VSI minimization. Maximum number of 200 iterations is considered as convergence criterion. The VSI converged on 0.07078 at 121st iteration. The time taken to achieve this convergence is 73 seconds. Figure 8.9 shows the comparison of L-index with different algorithms, and FPA-DE provides better results.

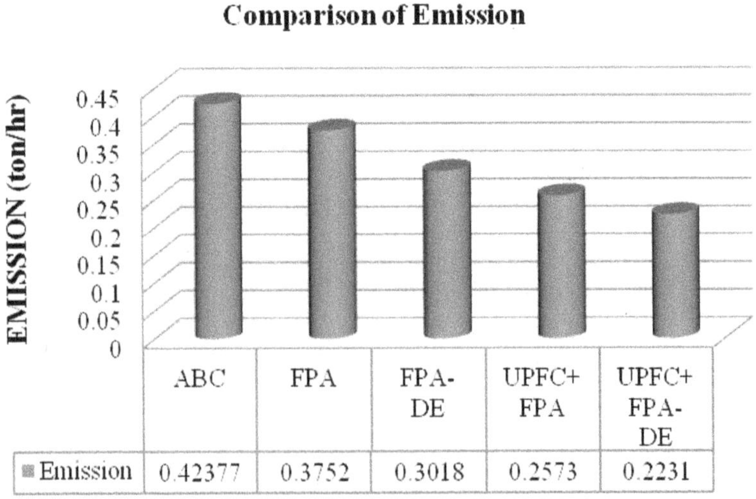

Comparison of Emission

	ABC	FPA	FPA-DE	UPFC+ FPA	UPFC+ FPA-DE
Emission	0.42377	0.3752	0.3018	0.2573	0.2231

FIGURE 8.5   Comparison of various algorithms versus emission cost.

TABLE 8.3

Loss Minimization Objective for IEEE 30 Bus

Generation	Without UPFC			With UPFC	
	ABC	FPA	FPA-DE	FPA	FPA-DE
V_{g1}	1.0227	1.009	0.979	0.981	0.954
V_{g2}	1.0035	1.001	0.980	0.981	0.964
V_{g3}	1.0252	0.973	0.970	0.938	0.920
V_{g4}	1.0076	0.976	0.980	0.968	0.964
V_{g5}	0.9821	1.020	0.976	0.944	0.925
V_{g6}	1.1000	0.977	0.951	0.975	1.018
T_1	1.0500	0.926	0.959	1.096	0.900
T_2	1.1000	0.993	0.951	1.047	1.100
T_3	0.9625	1.023	0.963	1.085	1.041
T_4	0.9000	1.079	1.016	0.905	0.983
UPFC-connected B/W buses	–	–	–	10–17	6–7
UPFC shunt power	–	–	–	$0 + j50.2579$	$0 + j33.706$
UPFC series power	–	–	–	$11.91 + j6.01$	$11.554 + j6.446$
Loss (MW)	14.0928	11.0975	10.3238	9.2670	5.4846

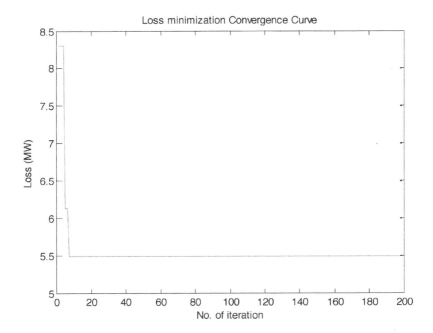

FIGURE 8.6 Loss convergence characteristics of hybrid FPA-DE with UPFC.

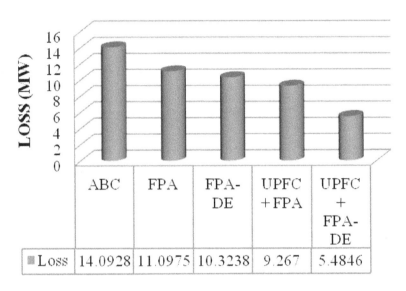

FIGURE 8.7   Comparison of various algorithms versus loss.

TABLE 8.4
VSI Objective for IEEE 30 Bus

Particulars	Without UPFC			With UPFC	
	ABC (Zhang and Li 2010)	FPA	FPA-DE	FPA	FPA-DE
V_{g1}	1.0227	1.009	1.049	1.024	1.023
V_{g2}	1.0035	1.001	1.010	1.010	1.041
V_{g3}	1.0252	0.973	1.001	1.002	1.037
V_{g4}	1.0076	0.976	1.008	1.014	1.037
V_{g5}	0.9821	1.020	0.974	1.008	0.970
V_{g6}	1.1000	0.977	1.015	1.016	0.985
UPFC-connected B/W buses	–	–	–	3–4	12–14
UPFC shunt power	–	–	–	$0 + j130.878$	$0 + j175.868$
UPFC series power	–	–	–	$0 + j4.4473$	$5.331 + j3.276$
L-index	0.1431	0.1320	0.0962	0.1146	0.07078

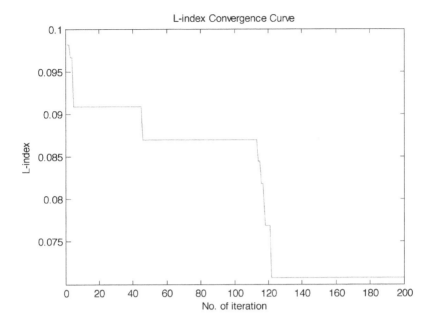

FIGURE 8.8   Voltage stability *L*-index convergence curve.

8.4.5   Pareto-Optimal Solution of Four Objectives

The Pareto technique is the important conspicuous candidate in solving MOOPFs, which works according to the introduced ideas of dominance. All Pareto-optimal solutions are sorted based on a decision-making strategy to choose the best compromise solution (BCS), which is one of the valid solutions in the Pareto front according to the user's tendencies. The four objective functions are minimized simultaneously, and the optimal results are given in the Table 8.5.

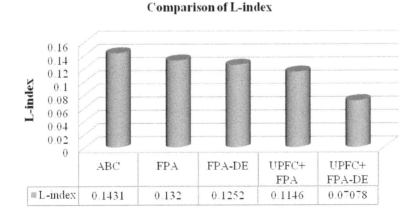

	ABC	FPA	FPA-DE	UPFC+ FPA	UPFC+ FPA-DE
L-index	0.1431	0.132	0.1252	0.1146	0.07078

FIGURE 8.9   Comparison of various algorithms versus VSI.

TABLE 8.5
Pareto Approach Solution for IEEE 30 Bus

Variables	ABC Algorithm	ABC FPA-DE	Cost FPA-DE	Cost FPA-DE	Emission FPA-DE	Emission FPA-DE	Loss FPA-DE	L-index FPA-DE	Multiobjective FPA-DE
P_{g1}	194.844	108.62	118.0496	147.37	86.815	173.36	61.957	59.052	78.430
P_{g2}	51.992	69.350	46.870	61.460	80.000	24.438	80.000	76.654	72.823
P_{g3}	15.000	23.801	40.149	17.305	50.000	22.142	44.081	41.990	36.936
P_{g4}	10.000	28.211	39.193	35.625	53.071	34.437	30.580	52.716	41.437
P_{g5}	10.000	23.064	30.000	19.003	30.000	19.375	23.165	26.536	11.170
P_{g6}	15.657	37.229	17.757	12.498	36.082	19.965	37.844	30.406	25.289
V_{g1}	1.0227	1.049	1.014	0.952	1.048	0.979	0.954	1.023	0.992
V_{g2}	1.0035	1.010	1.004	0.955	1.035	0.980	0.964	1.041	0.977
V_{g3}	1.0252	1.001	0.949	0.940	0.950	0.970	0.920	1.037	0.958
V_{g4}	1.0076	1.008	0.950	0.937	1.050	0.980	0.964	1.037	0.919
V_{g5}	0.9821	0.974	1.012	1.025	1.050	0.976	0.925	0.970	0.950
V_{g6}	1.1000	1.015	0.991	0.959	0.952	0.951	1.018	0.985	0.972
T_1	1.0500	1.033	1.040	0.945	1.075	0.959	0.900	0.955	1.098
T_2	1.1000	1.011	1.042	0.986	0.928	0.951	1.100	1.024	0.910
T_3	0.9625	1.002	1.002	0.922	1.091	0.963	1.041	0.978	1.100
T_4	0.9000	1.012	1.053	0.933	1.100	1.016	0.983	1.033	0.903
$UPFC_{Loc}$	—	—	6–7	—	6–10	—	6–7	12–14	6–7
$Shunt_{Pow}$	—	—	0+j233.668	—	0+j148.882	—	0+j33.706	0+j175.868	0+j4.8819
$Series_{Pow}$	—	—	2.163+j12.525	—	59.797+j4.459	—	11.554+j6.446	5.331+j3.276	0.5+j0.1775
Cost	945.4495	923.876	876.081	890.161	1179.83	895.112	965.761	996.805	870.316
Emission	0.42377	0.2419	0.2274	0.3018	0.2231	0.35348	0.21327	0.21201	0.2241
Loss	14.0928	6.8888	8.6186	9.8666	8.57827	10.3238	5.4846	4.8247	5.4931
VSI	0.1431	0.0962	0.1068	0.1363	0.14657	0.12519	0.09341	.07078	0.11626

8.5 CASE STUDY II: SOLVING THE PROBLEM ON THE IEEE 57 BUS TEST SYSTEM

The IEEE 57 bus system has 80 transmission lines; 7 generation units at buses 1, 2, 3, 6, 8, 9, and 12; 3 compensator components at buses 18, 25, and 53; and 15 transformers. The entire load demand of this network is 1250.8 MW, and bus 1 is the slack bus. The acceptable range of voltage magnitude for both controlled buses (PV) and (PQ) load buses is set to [0.95–1.05] pu. The IEEE 57 bus single-line diagram is shown in Figure 8.10.

The hybrid FPA-DE algorithm is implemented for load demand of 1250.8 MW and 336.4 MVAR, thereby reducing the generators cost and the results are compared with MATLAB Interior Point (MIP) Solver (conventional mathematical method) and ABC algorithm (intelligent algorithm), shown in Table 8.6. Upon introduction of UPFC device in the system, it takes up some real power and reactive power load,

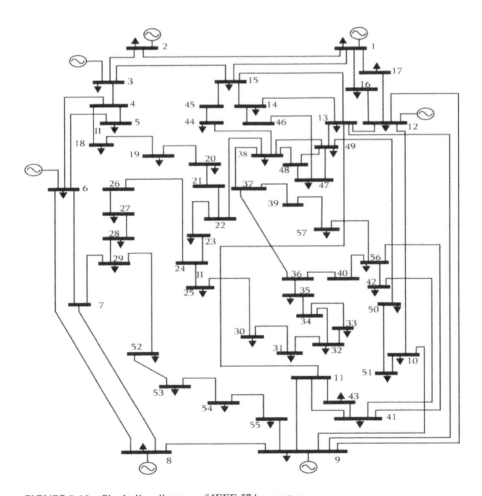

FIGURE 8.10   Single-line diagram of IEEE 57 bus system.

TABLE 8.6

Generation Cost Optimization – IEEE 57 Bus

Control Variables	Without UPFC		With UPFC	
	MIP	ABC Rezaei Adaryani et al. (2013)	FPA	FPA-DE
P_{g1} (MW)	142.63	142.8106	133.1657	123.61
P_{g2} (MW)	87.82	90.0328	87.8200	87.82
Pg_3 (MW)	45.07	44.5147	45.0700	45.07
P_{g4} (MW)	72.90	74.2003	72.9000	72.9
P_{g5} (MW)	459.83	454.8475	459.8300	459.83
P_{g6} (MW)	97.51	96.8847	97.5100	97.51
P_{g7} (MW)	361.54	362.7722	361.5400	361.54
V_{g1} (pu)	1.009	1.0423	1.04	1.04
V_{g2} (pu)	1.008	1.0411	1.01	1.01
V_{g3} (pu)	1.003	1.0385	0.985	0.985
V_{g4} (pu)	1.026	1.0549	0.98	0.980
V_{g5} (pu)	1.044	1.0640	1.005	1.005
V_{g6} (pu)	1.004	1.0369	0.98	0.980
V_{g7} (pu)	0.992	1.0406	1.015	1.015
T_1 (pu)	0.9700	0.9375	0.9033	1.0006
T_2 (pu)	0.9780	1.0500	0.9984	1.0158
T_3 (pu)	1.0430	0.9750	0.9821	1.0940
T_4 (pu)	1.0000	0.9500	1.0594	1.0511
T_5 (pu)	1.0000	1.0125	1.0650	0.9676
T_6 (pu)	1.0430	1.0000	0.9809	0.9463
T_7 (pu)	0.9670	1.0125	1.0662	0.9687
T_8 (pu)	0.9750	0.9125	1.0028	0.9566
T_9 (pu)	0.9550	0.9000	1.0237	0.9342
T_{10} (pu)	0.9550	1.0125	0.9266	0.9272
T_{11} (pu)	0.9000	0.9875	0.9980	0.9117
T_{12} (pu)	0.9300	1.0000	0.9547	1.0225
T_{13} (pu)	0.8950	0.9625	1.0818	0.9690
T_{14} (pu)	0.9580	0.9625	0.9029	0.9285
T_{15} (pu)	0.9580	0.9625	0.9228	1.0288
T_{16} (pu)	0.9800	0.9250	0.9728	1.0226
T_{17} (pu)	0.9400	0.9875	1.0061	1.0557
Q_{c1} (MVAR)	10	16	95.6474	63.4975
Q_{c2} (MVAR)	5.9	15	17.4762	15.3182
Q_{c3} (MVAR)	6.3	14	27.2594	74.8808
$UPFC_{Loc}$	–	–	10–51	44–45
$Shunt_{Pow}$	–	–	$0 + j54.4338$	$0 + j19.3683$
$Series_{Pow}$	–	–	$25.0791 + j111.749$	$32.3597 + j82.1761$
Cost ($/hour)	41737.79	41693.9589	41345.4518	40963.9832

and thereby the generating cost is reduced. Compared with FPA, hybrid FPA-DE algorithm gives minimum generating cost, as given in the last column of Table 8.6.

8.6 CONCLUSION

This chapter clearly explains about the hybrid FPA-DE algorithm for solving the MOOPF problem with flexible alternating current transmission system (FACTS) devices. The developed algorithm and flowchart are implemented using MATLAB and their optimal results are presented. Best results were obtained from the hybrid FPA-DE with respect to other alternative algorithms. A comprehensive performance of this FPA-DE approach is determined using IEEE 30 and 57 bus systems. The potential outcome of the hybrid FPA-DE includes minimum cost, generators less emission, and reduced loss and improves VSI for the objective function with and without valve point loading. For IEEE 30 bus system, in comparison with previous literature, the generating cost is reduced by 7.95%, emission is reduced by 47.12%, loss is reduced by 61.02%, and stability index is improved by 18.76%. This shows the superiority of the hybrid FPA-DE in solving MOOPF problem.

8.7 SUMMARY

i. A straightforward hybrid optimization algorithm to solve different versions of OPF is proposed.

ii. Two IEEE standard test systems are studied to scrutinize the strong presentation of the algorithm in dealing with the most practical OPF problem with different constraints.

iii. The FPA-DE approach is very powerful and robust for answering the OPF problem in all case studies regardless of the problems, which has been proved to be better optimization algorithm.

iv. Contrary to other algorithms, the obtained execution times from the proposed approach show that the FPA-DE algorithm managed to converge to an optimal solution in less execution time than the other algorithms for all of the defined case studies.

REFERENCES

Abido, MA & Al-Ali, NA, 2009, 'Multi-objective differential evolution for optimal power flow', Proceedings of the Second IEEE International Conference on Power Engineering, Energy and Electrical Drives, IEEE, pp. 101–106.

Bhattacharya, A & Chattopadhyay PK, 2010, 'Hybrid differential evolution with biogeography-based optimization for solution of economic load dispatch', IEEE Transactions on Power Systems, vol. 25, no. 4, pp. 1955–1964.

Chung, CY, 2010, 'Hybrid algorithm of differential evolution and evolutionary programming optimal reactive power flow', IET Generation Transmission and Distribution, vol. 4, no. 1, pp. 84–93.

Hariharan, T & Mohana Sundaram, K, 2016, 'A novel hybrid FPA-DE algorithm for solving multiobjective optimal power flow with unified power flow controller', Journal of Computational and Theoretical Nanoscience, vol. 13, pp. 5199–5208.

Herbadji, O, Slimani, L & Bouktir, T, 2016, 'Solving bi-objective optimal power flow using hybrid method of biogeography-based optimization and differential evolution algorithm: A case study of the Algerian electrical network', Journal of Electrical Systems, vol. 12, no. 1, pp. 197–215.

Li, C, Zhao, H & Chen, T, 2010, 'The hybrid differential evolution algorithm for optimal power flow based on simulated annealing and tabu search', Proceedings of the 4th IEEE International Conference on Management and Service Science, IEEE, pp. 1–7.

Ponnin Thilagar, P & Harikrishnan, R, 2015, 'Application of intelligent firefly algorithm to solve OPF with STATCOM', Indian Journal of Science and Technology, vol. 8, no. 22, pp. 1–5.

Prathiba, R, Moses, MB & Sakthivel, S, 2014, 'Flower pollination algorithm applied for different economic load dispatch problems', International Journal of Engineering and Technology, vol. 6, no. 2, pp. 1009–1016.

Sahu, RK Panda, S & Padhan, S, 2015, 'A hybrid firefly algorithm and pattern search technique for automatic generation control of multi area power systems', International Journal of Electrical Power & Energy Systems, vol. 64, pp. 9–23.

Shilaja, C & Ravi, K, 2016 'Optimal line flow in conventional power system using Euclidean affine flower pollination algorithm', International Journal of Renewable Energy Research, vol. 6, no. 1, pp. 335–342.

Yan, H & Li, X, 2010, 'Stochastic optimal power flow based improved differential evolution', Conference Proceedings: 2010 Second WRI Global Congress on Intelligent Systems, IEEE, vol. 3, pp. 243–246.

Zhang, H & Li, P, 2010, 'Probabilistic analysis for optimal power flow under uncertainty', IET Generation, Transmission and Distribution, vol. 4, no. 5, pp. 553–561.

APPENDIX I

A.1 IEEE 30 BUS DATA

TABLE A.1
Generator and Cost Coefficients Data

Generator	P_{min} (MW)	P_{max} (MW)	α ($/hour)	β ($/MWhr)	γ ($/MW²hr)
G_1	50	200	0.00375	2	0
G_2	20	80	0.0175	1.75	0
G_3	15	50	0.0625	1	0
G_4	10	35	0.0083	3.25	0
G_5	10	30	0.025	3	0
G_6	12	40	0.025	3	0

TABLE A.2
Cost and Emission Coefficients of IEEE 30 Bus System

Coefficients	G_1	G_2	G_3	G_4	G_5	G_6
Fuel cost coefficents						
A	100	120	40	60	40	100
B	200	150	180	100	180	150
C	10	10	20	10	20	10
D	15	10	10	5	5	5
E	6.28	8.98	14.78	20.94	25.13	18.48
Emission cost coefficients						
Υ	0.06490	0.05638	0.04586	0.0338	0.04586	0.05151
B	−0.5554	−0.06047	−0.05094	−0.0355	−0.05094	−0.05555
A	0.04091	0.02543	0.04258	0.05326	0.04258	0.06131
Ξ	0.0002	0.0005	0.000001	0.002	0.000001	0.00001
Λ	2.857	3.333	8.00	2.00	8.00	6.667

TABLE A.3
IEEE 30 Bus Data

S. No.	Bus Type	V_m (pu)	V_{ang} (pu)	P_L (MW)	Q_L (MVAR)	Q_{min} (MVAR)	Q_{max} (MVAR)	Injected MVAR
1	Slack	1.05	0	0	0	0	0	0
2	PV	1.01	0	21.7	12.7	−40	50	0
3	PQ	1	0	2.4	1.2	0	0	0
4	PQ	1.01	0	7.6	1.6	0	0	0
5	PV	1.01	0	94.2	19	−40	40	0
6	PQ	1	0	0	0	0	0	0
7	PQ	1	0	22.8	10.9	0	0	0
8	PV	1.01	0	30	30	−10	40	0
9	PQ	1	0	0	0	0	0	0
10	PQ	1	0	5.8	2	0	0	19
11	PV	1.01	0	0	0	−6	24	0
12	PQ	1	0	11.2	7.5	0	0	0
13	PV	1.01	0	0	0	−6	24	0
14	PQ	1	0	6.2	1.6	0	0	0
15	PQ	1	0	8.2	2.5	0	0	0
16	PQ	1	0	3.5	1.8	0	0	0
17	PQ	1	0	9	5.8	0	0	0
18	PQ	1	0	3.2	0.9	0	0	0
19	PQ	1	0	9.5	3.4	0	0	0
20	PQ	1	0	2.2	0.7	0	0	0
21	PQ	1	0	17.5	11.2	0	0	0
22	PQ	1	0	0	0	0	0	0
23	PQ	1	0	3.2	1.6	0	0	0
24	PQ	1	0	8.7	6.7	0	0	4.3
25	PQ	1	0	0	0	0	0	0
26	PQ	1	0	3.5	2.3	0	0	0
27	PQ	1	0	0	0	0	0	0
28	PQ	1	0	0	0	0	0	0
29	PQ	1	0	2.4	0.9	0	0	0
30	PQ	1	0	10.6	1.9	0	0	0

TABLE A.4
IEEE 30 Bus Line Data

From Bus	To Bus	R (pu)	X (pu)	B/2 (pu)	Type
1	2	0.0192	0.0575	0.0264	Line
1	3	0.0452	0.1852	0.0204	Line
2	4	0.057	0.1737	0.0184	Line
2	5	0.0472	0.1983	0.0209	Line
2	6	0.0581	0.1763	0.0187	Line
3	4	0.0132	0.0379	0.0042	Line
4	6	0.0119	0.0414	0.0045	Line
4	12	0	0.256	0	Transformer
5	7	0.046	0.116	0.0102	Line
6	7	0.0267	0.082	0.0085	Line
6	8	0.012	0.042	0.0045	Line
6	9	0	0.208	0	Transformer
6	10	0	0.556	0	Transformer
6	28	0.0169	0.0599	0.0065	Line
8	28	0.0636	0.2	0.0214	Line
9	11	0	0.208	0	Line
9	10	0	0.11	0	Line
10	20	0.0936	0.209	0	Line
10	17	0.0324	0.0845	0	Line
10	21	0.0348	0.0749	0	Line
10	22	0.0727	0.1499	0	Line
12	13	0	0.14	0	Line
12	14	0.1231	0.2559	0	Line
12	15	0.0662	0.1304	0	Line
12	16	0.0945	0.1987	0	Line
14	15	0.221	0.1997	0	Line
15	18	0.107	0.2185	0	Line
15	23	0.1	0.202	0	Line
16	17	0.0824	0.1932	0	Line
18	19	0.0639	0.1292	0	Line
19	20	0.034	0.068	0	Line
21	22	0.0116	0.0236	0	Line
22	24	0.115	0.179	0	Line
23	24	0.132	0.27	0	Line
24	25	0.1885	0.3292	0	Line
25	26	0.2544	0.38	0	Line
25	27	0.1093	0.2087	0	Line
27	29	0.2198	0.4153	0	Line
27	30	0.3202	0.6027	0	Line
28	27	0	0.396	0	Transformer
29	30	0.2399	0.4533	0	Line

A.2 IEEE 57 Bus data

TABLE A.5

Line Data of IEEE 57 Bus System

From Bus	To Bus	R	X	Y_{cp}	Y_{cq}	Tap Ratio
1	2	0.0083	0.0280	0.0645	0.0645	1
2	3	0.0298	0.0850	0.0409	0.0409	1
3	4	0.0112	0.0366	0.0190	0.0190	1
4	5	0.0625	0.1320	0.0129	0.0129	1
4	6	0.0430	0.1480	0.0174	0.0174	1
6	7	0.0200	0.1020	0.0138	0.0138	1
6	8	0.0399	0.1730	0.0235	0.0235	1
8	9	0.0099	0.0505	0.0274	0.0274	1
9	10	0.0369	0.1679	0.0220	0.0220	1
9	11	0.0258	0.0848	0.0109	0.0109	1
9	12	0.0648	0.2950	0.0386	0.0386	1
9	13	0.0481	0.1580	0.0203	0.0203	1
13	14	0.0132	0.0434	0.0055	0.0055	1
13	15	0.0269	0.0869	0.0115	0.0115	1
1	15	0.0178	0.0910	0.0494	0.0494	1
1	16	0.0454	0.2060	0.0273	0.0273	1
1	17	0.0238	0.1080	0.0143	0.0143	1
3	15	0.0163	0.053	0.0272	0.0272	1
4	18	0	0.555	0	0	0.97
4	18	0	0.430	0	0	0.978
5	6	0.0302	0.0641	0.0062	0.0062	1
7	8	0.0139	0.0712	0.0097	0.0097	1
10	12	0.0277	0.1262	0.0164	0.0164	1
11	13	0.0223	0.0732	0.0094	0.0094	1
12	13	0.0178	0.0580	0.0302	0.0302	1
12	16	0.0180	0.0813	0.0108	0.0108	1
12	17	0.0397	0.1790	0.0238	0.0238	1
14	15	0.0171	0.0547	0.0074	0.0074	1
18	19	0.4610	0.6850	0	0	1
19	20	0.2830	0.4340	0	0	1
20	21	0	0.7767	0	0	1.043
21	22	0.0736	0.1170	0	0	1
22	23	0.0099	0.0152	0	0	1
23	24	0.1660	0.2560	0.0042	0.0042	1
24	25	0	1.182	0	0	1.01
24	25	0	1.230	0	0	1.01
24	26	0	0.0473	0	0	1.043
26	27	0.165	0.2540	0	0	1
27	28	0.0618	0.0954	0	0	1

TABLE A.5 (CONTINUED)
Line Data of IEEE 57 Bus System

From Bus	To Bus	R	X	Y_{cp}	Y_{cq}	Tap Ratio
28	29	0.0418	0.0587	0	0	1
7	29	0	0.0648	0	0	0.967
25	30	0.135	0.202	0	0	1
30	31	0.326	0.497	0	0	1
31	32	0.507	0.755	0	0	1
32	33	0.0392	0.0360	0	0	1
32	34	0	0.953	0	0	0.975
34	35	0.052	0.0780	0.0016	0.0016	1
35	36	0.043	0.0537	0.0008	0.0008	1
36	37	0.0290	0.0366	0	0	1
37	38	0.0651	0.1009	0.0010	0.0010	1
37	39	0.0239	0.0379	0	0	1
36	40	0.0300	0.0466	0	0	1
22	38	0.0192	0.0295	0	0	1
11	41	0	0.7490	0	0	0.955
41	42	0.207	0.3520	0	0	1
41	43	0	0.412	0	0	1
38	44	0.0289	0.0585	0.0010	0.0010	1
15	45	0	0.1042	0	0	0.955
14	46	0	0.0735	0	0	0.9
46	47	0.023	0.068	0.0016	0.0016	1
47	48	0.018	0.0233	0	0	1
48	49	0.083	0.1290	0.0024	0.0024	1
49	50	0.0801	0.1280	0	0	1
50	51	0.1386	0.2200	0	0	1
10	51	0	0.0712	0	0	0.93
13	49	0	0.1910	0	0	0.895
29	52	0.1442	0.1870	0	0	1
52	53	0.0762	0.0984	0	0	1
53	54	0.1878	0.2320	0	0	1
54	55	0.1732	0.2265	0	0	1
11	43	0	0.1530	0	0	0.958
44	45	0.0624	0.0020	0.0020	0.0020	1
40	56	0	1.1950	0	0	0.956
56	41	0.5530	0.5490	0	0	1
56	42	0.2125	0.3540	0	0	1
39	57	0	1.3550	0	0	0.98
57	56	0.1740	0.2600	0	0	1
38	49	0.1150	0.1770	0.0030	0.0030	1
38	48	0.0312	0.0482	0	0	1
9	55	0	0.1205	0	0	0.94

TABLE A.6
Bus Data of IEEE 57 Bus System

Bus	V_{spec}	Type	P_{gen}	Q_{gen}	P_{load}	Q_{load}
1	1.040000	Slack	0.000000	0.000000	0.550000	0.170000
2	1.010000	PV	0.000000	0.000000	0.030000	0.880000
3	0.985000	PV	0.400000	0.000000	0.410000	0.210000
4	0.980000	PV	0.000000	0.000000	0.750000	0.020000
5	1.005000	PV	4.500000	0.000000	1.500000	0.220000
6	0.980000	PV	0.000000	0.000000	1.210000	0.260000
7	1.015000	PV	3.100000	0.000000	3.770000	0.240000
8	1.000000	PQ	0.000000	0.000000	0.000000	0.000000
9	1.000000	PQ	0.000000	0.000000	0.130000	0.040000
10	1.000000	PQ	0.000000	0.000000	0.000000	0.000000
11	1.000000	PQ	0.000000	0.000000	0.050000	0.020000
12	1.000000	PQ	0.000000	0.000000	0.000000	0.000000
13	1.000000	PQ	0.000000	0.000000	0.180000	0.023000
14	1.000000	PQ	0.000000	0.000000	0.105000	0.053000
15	1.000000	PQ	0.000000	0.000000	0.220000	0.050000
16	1.000000	PQ	0.000000	0.000000	0.430000	0.030000
17	1.000000	PQ	0.000000	0.000000	0.420000	0.080000
18	1.000000	PQ	0.000000	0.000000	0.272000	0.098000
19	1.000000	PQ	0.000000	0.000000	0.033000	0.006000
20	1.000000	PQ	0.000000	0.000000	0.023000	0.001000
21	1.000000	PQ	0.000000	0.000000	0.000000	0.000000
22	1.000000	PQ	0.000000	0.000000	0.000000	0.000000
23	1.000000	PQ	0.000000	0.000000	0.063000	0.002100
24	1.000000	PQ	0.000000	0.000000	0.000000	0.000000
25	1.000000	PQ	0.000000	0.000000	0.063000	0.003200
26	1.000000	PQ	0.000000	0.000000	0.000000	0.000000
27	1.000000	PQ	0.000000	0.000000	0.093000	0.005000
28	1.000000	PQ	0.000000	0.000000	0.046000	0.002300
29	1.000000	PQ	0.000000	0.000000	0.170000	0.002600
30	1.000000	PQ	0.000000	0.000000	0.036000	0.001800
31	1.000000	PQ	0.000000	0.000000	0.058000	0.002900
32	1.000000	PQ	0.000000	0.000000	0.016000	0.008000
33	1.000000	PQ	0.000000	0.000000	0.038000	0.019000
34	1.000000	PQ	0.000000	0.000000	0.000000	0.000000
35	1.000000	PQ	0.000000	0.000000	0.060000	0.030000
36	1.000000	PQ	0.000000	0.000000	0.000000	0.000000
37	1.000000	PQ	0.000000	0.000000	0.000000	0.000000
38	1.000000	PQ	0.000000	0.000000	0.140000	0.070000
39	1.000000	PQ	0.000000	0.000000	0.000000	0.000000
40	1.000000	PQ	0.000000	0.000000	0.000000	0.000000
41	1.000000	PQ	0.000000	0.000000	0.063000	0.030000
42	1.000000	PQ	0.000000	0.000000	0.071000	0.040000
43	1.000000	PQ	0.000000	0.000000	0.020000	0.010000

TABLE A.6 (CONTINUED)
Bus Data of IEEE 57 Bus System

Bus	V_{spec}	Type	P_{gen}	Q_{gen}	P_{load}	Q_{load}
44	1.000000	PQ	0.000000	0.000000	0.120000	0.018000
45	1.000000	PQ	0.000000	0.000000	0.000000	0.000000
46	1.000000	PQ	0.000000	0.000000	0.000000	0.000000
47	1.000000	PQ	0.000000	0.000000	0.297000	0.116000
48	1.000000	PQ	0.000000	0.000000	0.000000	0.000000
49	1.000000	PQ	0.000000	0.000000	0.180000	0.085000
50	1.000000	PQ	0.000000	0.000000	0.210000	0.105000
51	1.000000	PQ	0.000000	0.000000	0.180000	0.053000
52	1.000000	PQ	0.000000	0.000000	0.049000	0.022000
53	1.000000	PQ	0.000000	0.000000	0.200000	0.100000
54	1.000000	PQ	0.000000	0.000000	0.041000	0.014000
55	1.000000	PQ	0.000000	0.000000	0.068000	0.034000
56	1.000000	PQ	0.000000	0.000000	0.076000	0.022000
57	1.000000	PQ	0.000000	0.000000	0.067000	0.020000

TABLE A.7
Generator Data of IEEE 57 Bus System

Bus No.	P_G^{min} (MW)	P_G^{max} (MW)	Q_G^{min} (MVAR)	Q_G^{max} (MVAR)
1	0.2	0.5	0.0	0.0
2	0.15	0.9	0.5	−0.17
3	0.1	5	0.6	−0.1
4	0.1	0.5	0.25	−0.08
5	0.12	0.5	2	−1.4
6	0.1	3.6	0.09	−0.03
7	0.5	5.5	1.55	−0.5

TABLE A.8
Cost and Emission Coefficients of IEEE 57 Bus System

Unit	a_i	b_i	c_i	α_i	β_i	γ_i
1	115	2.00	0.0055	21.022	−1.820	0.031
2	40	3.50	0.0060	22.050	−1.249	0.013
3	122	3.15	0.0050	22.983	−1.355	0.012
4	125	3.05	0.0050	21.313	−1.900	0.020
5	120	2.75	0.0070	21.900	0.805	0.007
6	70	3.45	0.0070	23.001	−1.401	0.015
7	150	1.89	0.0050	23.333	−1.500	0.016

9 SSSC and TCSC with Fuzzy Logic Controller for Damping SSR Oscillations

LEARNING OUTCOME

i. To study about the IEEE Second Benchmark Model (SBM) system.
ii. To study about static synchronous series compensator (SSSC) and thyristor-controlled series capacitor (TCSC) with a conventional proportional-integral (PI)controller for subsynchronous resonance (SSR)oscillation damping under steady state and transient state.
iii. To study about SSSC and TCSC with a fuzzy logic controller (FLC) for SSR oscillation damping under steady state and transient state.
iv. To analyze the system with different load in an IEEE SBM system.

9.1   INTRODUCTION

The impact of heavy disturbances introduced in the power system operation will result in a nonlinear behavior, and the controllers fail to do the necessary damping of oscillations. A high destabilizing impact may lead to successive disturbance in the controller, thereby inserting negative damping. To avoid such situations, the dynamic nature of the system network has to be considered by the controlling scheme. The development of fuzzy logic controller (FLC) provides a diversified way to control the nonlinear process. It is general formulation that a well-defined fuzzy controller can provide a better performance in the presence of variations in system operating parameters, load, and external disturbances. Varma et al (2006) presented a method of mitigating SSR through PMU-acquired remote generator speed. In the same year, he also proposed a static var compensator for series compensation of a wind farm. Kumar et al (2011) has presented a detailed comparison of auxiliary signals for SSR mitigation through TCR-FC. Hosseini et al (2013) has proven that Fuzzy Logic Based Damping Controller (FLBDC) with flexible alternating current transmission system (FACTS) device improves power system operation in a wide range of operating conditions. A proportional-integral (PI) conventional controller in static synchronous series compensator (SSSC) and thyristor-controlled series capacitor (TCSC) is also used for controlling subsynchronous resonance (SSR) oscillations.

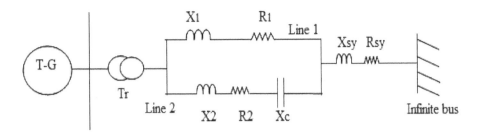

FIGURE 9.1   An IEEE SBM system.

Sindhu Thampatty et al (2011) has also proposed a method which Amini et al (2013) has proven for the reduction of subsynchronous resonance through artificial intelligence. Vivek et al (2014) proposed a topology for reduction of SSR in a series-compensated wind generation system. Li et al (2016) has presented the impact of wind power on subsynchronous resonance of turbine-generator units. It has been proven that wind power generation influences SSR to a great extent. Zhu et al (2018) proposed a topology for mitigation of SSR for a power system with unified power flow controller (UPFC). The conventional control method requires system mathematical model and the system behavior is not satisfactory due to parametric variations, but FLC-based system removes such complications and ends in efficient SSR oscillation damping. The researchers as shown in the previous literature survey have carried out extensive work on the control of various parameters by using Mamdani-based FLC. This chapter explains about the importance of PI controller in controlling the SSR oscillations and also a sophisticated controller called the Mamdani-based FLC implemented in an IEEE SBM system with SSSC and TCSC.

9.2   IEEE SBM SYSTEM

The IEEE SBM system is shown in Figure 9.1. The model has a single machine connected to an infinite bus power system through parallel AC transmission lines, one of which is connected to a series capacitor with reactance X_c.

The mechanical part as shown in Figure 9.2 consists of a four springmass system: high-pressure turbine (HP), low-pressure turbine (LP), generator (GEN), and the exciter (EXC) coupled to a common shaft.

There are several analytical tools in order to study SSR. Frequency scanning and eigenvalue and time domain analyses are the most common tools used to evaluate SSR and its effects.

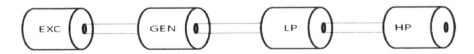

FIGURE 9.2   Mechanical parts of the generating unit.

9.3 SSSC WITH A CONVENTIONAL PI CONTROLLER FOR SSR OSCILLATION DAMPING

An IEEE SBM system connected with a series-compensated SSSC in Line 1 is shown in Figure 9.3. The SSSC serves the purpose of injecting a voltage V_{SSSC} in quadrature to line current; thus, line reactance is varied so as to damp SSR-related oscillations. A conventional PI controller in SSSC will damp SSR oscillations at generator turbine shaft as well as rotor speed oscillations. Rotor speed deviation ($\Delta\omega$) of generator that is practically measurable in real-time application is used as the input signal for the PI-controlled SSSC.

The PI controller operates in a wide range of series compensation ratios and can improve oscillation damping. The implemented control scheme of a PI auxiliary controller with SSSC is shown in Figure 9.4. The output of the phase-locked loop (PLL) θ (transmission line current phase angle) is used to measure the direct axis (V_d) and quadrature axis (V_q) components of the AC three-phase voltage and currents I_d, I_q, respectively.

In order to control V_q and V_d, a conventional PI controller is used as auxiliary controller or subsynchronous damping controller (SSDC). The principal strategy in controlling compensators for damping SSR oscillations is selection of stabilizing signals. In this controller, rotor speed deviation ($\Delta\omega$) of generator is used as the stabilizing signal. With respect to schematic diagram, the SSDC uses rotor speed deviation of generator as the input signal. The input signal after passing through PI controller is used to modulate the SSSC-injected voltage V_q to improve the damping of the unstable torsional oscillatory modes. V_{qref} denotes the injected reference voltage as desired by the control loop. The value of obtained V_{dcnv} and V_{qcnv} replicates the components of converter voltage (V_{conv}), which is in phase and in quadrature with line current, respectively. The control system implemented here consists of measurement systems that measure the q components of AC positive sequence of voltages V_1 and V_2 (V_{1q} and V_{2q}) as well as the DC voltage V_{dc}. The AC and DC voltage regulators here

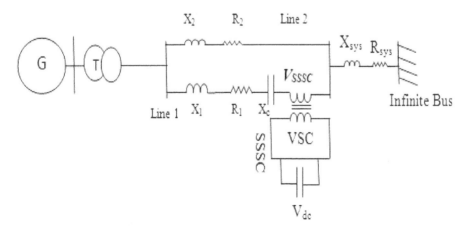

FIGURE 9.3　An IEEE SBM system with SSSC.

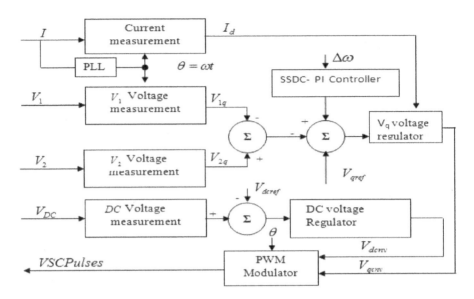

FIGURE 9.4   Block diagram of SSSC with PI controller.

compute the two components of the converter voltage (V_{dcnv}) and (V_{qcnv}), which are needed for obtaining the desired DC voltage ($V_{dc\ ref}$) and the injected voltage (V_{qref}). The V_q voltage regulator is supported by a feed forward type regulator, which helps in predicting the voltage V_{cnv} that is injected on the voltage source converter (VSC) side of the transformer from the measured current I_d. The PI controller parameters are fine-tuned by hit and trial method.

9.4   TCSC WITH A CONVENTIONAL PI CONTROLLER FOR SSR OSCILLATION DAMPING

A TCSC can be operated in different control methodologies to damp SSR oscillations. The robustness of constant current control along with a conventional PI controller for SSR oscillation damping is studied. The analysis is conducted for various level of series compensation with constant current control methodology as described below.

A TCSC device operating in local current control is used to substitute a part of the fixed series compensation in an IEEE SBM system. A closed-loop current control along with PI controller is shown in Figure 9.5.

The I_{ref} is the precontingency current. The main current controller is a PI controller whose parameters are altered through transient simulation studies by hit and trial method to attain minimum settling time. The value of K_p and K_I are 10.16 and 3.47, respectively.

A classical typical TCSC constant current (CC) controller model is shown in Figure 9.6 where the desired value of line current is taken as the reference signal. The desired line-current amplitude is given as a reference signal to the TCSC

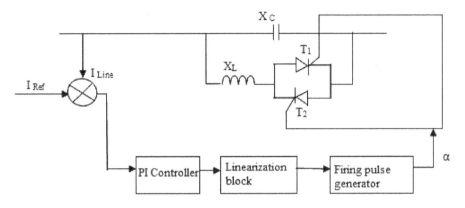

FIGURE 9.5   Block diagram of a conventional PI controller with TCSC.

controller, which strives to retain the actual line current at this value. The 3-phase current is measured, rectified, and then passed through a filtering block. The controller is typically of the PI type that provides the desired susceptance signal within the preset limits. A linearizer block converts the susceptance signal into a firing-angle signal.

For validating the PI controller's performance, the system response is compared with the following cases: (i) with PI controller based SSSC; (ii) with PI controller based TCSC. These two cases are carried out under different operating states:50%, 60%, 70%, and 80% series compensation during steady state and transient state.

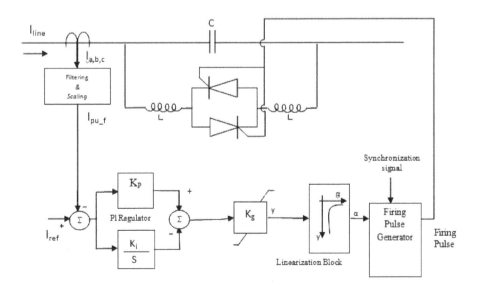

FIGURE 9.6   TCSC constant current control.

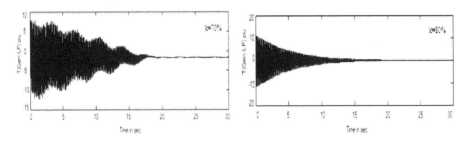

FIGURE 9.7 *T*(Gen-LP)pu during steady state at 70% and 80% series compensation with PI-controlled SSSC.

9.5 STEADY-STATE SSR ANALYSIS IN STEAM TURBINE GENERATOR WITH CONTROLLERS

The SSR oscillations were observed during steady-state condition at 70% and 80% series compensation levels *K*, and hence the system is tested with PI-controlled SSSC and TCSC.

CASE I TORQUE BETWEEN GENERATOR AND LOW-PRESSURE TURBINE
T(GEN-LP)PU DURING STEADY STATE WITH PI-CONTROLLED
SSSC AND TCSC IN AN IEEE SBM SYSTEM

Based on Figure 9.7, with SSSC controller, the Gen-LP torque oscillation at $K = 70\%$ reaches a maximum value of 8pu and, due to the effect of controller, the oscillation gets damped at 17 seconds. At $K = 80\%$, Gen-LP torque oscillation occurs with a magnitude of 10 pu And takes 20 seconds to get completely damped.

Figure 9.8 shows *T*(Gen-LP)pu during steady state at 70% and 80% series compensation with PI-controlled TCSC. At $K = 70\%$, the oscillations reach a maximum value of 10 pu, and due to the effect of TCSC controller, the oscillation gets completely damped at 19 seconds. At $K = 80\%$, Gen-LP torque oscillation attains the magnitude of 12 pu, and it takes 24 seconds to get completely damped.

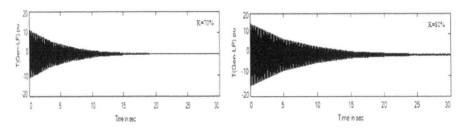

FIGURE 9.8 *T*(Gen-LP)pu during steady state at 70% and 80% series compensation with PI-controlled TCSC.

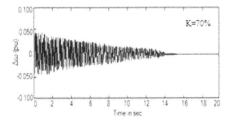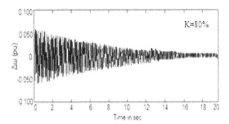

FIGURE 9.9   Rotor speed deviation ($\Delta\omega$) during steady state at 70% and 80% series compensation with PI-controlled SSSC.

CASE II ROTOR SPEED DEVIATION ($\Delta\Omega$)DURING STEADY STATE WITH PI-CONTROLLED SSSC AND TCSC IN AN IEEE SBM SYSTEM

Figure 9.9 shows the rotor speed deviation ($\Delta\omega$) during steady state at 70% and 80% series compensation with PI-controlled SSSC. From Figure 9.9, it is observed that the rotor speed deviation ($\Delta\omega$) observed tends to reach a maximum value of 0.040 pu at $K = 70\%$ and 0.052 pu at $K = 80\%$, and due to the impact of SSSC controller, the oscillation gets damped at 15 and 20 seconds, respectively.

Figure 9.10 shows the rotor speed deviation ($\Delta\omega$) during steady state at 70% and 80% series compensation with PI-controlled TCSC. It can be observed from Figure 9.10 that rotor speed deviation ($\Delta\omega$) reaches a maximum value of 0.050 pu at $K = 70\%$ and 0.055 pu at $K = 80\%$. Due to the impact of TCSC controller, the oscillation gets damped at 20 and 20 seconds, respectively.

9.6   TRANSIENT-STATE SSR EFFECT ANALYSIS IN STEAM TURBINE GENERATOR WITH CONTROLLERS

In order to demonstrate the influence of PI-controlled SSSC and TCSC on SSR oscillations effects damping during transient state, simulation runs were done with different levels of series compensation percentage K.

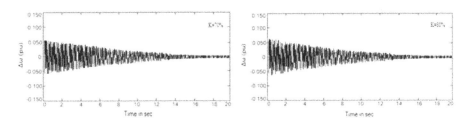

FIGURE 9.10   Rotor speed deviation ($\Delta\omega$) during steady state at 70% and 80% series compensation with PI-controlled TCSC.

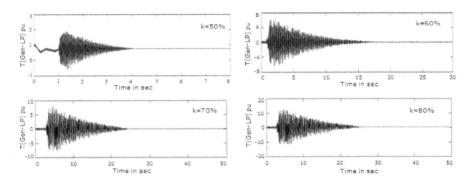

FIGURE 9.11 *T*(Gen-LP)pu during transient state at 50%, 60%, 70%, and 80% series compensation with PI-controlled SSSC.

CASE I TORQUE BETWEEN GENERATOR AND LOW-PRESSURE TURBINE *T*(GEN-LP) PU DURING TRANSIENT STATE WITH PI-CONTROLLED SSSC AND TCSC IN AN IEEE SBM SYSTEM

The plots of torque between generator and low-pressure turbine for different value *K* with respect to settling time is given in Figure 9.11 All the plots are simulated results with MATLAB software for a PI-SSSC-controlled IEEESBM system. It can be observed from Figure 9.11 that due to the influence of proposed controller, the damping of turbine-generator oscillations is achieved at all levels of series compensation. But one important observation made is that as *K* value increases from 50% to 80%, *T*(Gen-LP) magnitude also increases, thereby increasing the settling time of oscillatory torque.

The effect of PI-controlled TCSC in damping *T*(Gen-LP) is shown in Figure 9.12. At *K* = 50% of series compensation, it could be observed that the value of Generator-LP torque reaches a maximum of 2 pu and gets damped completely at

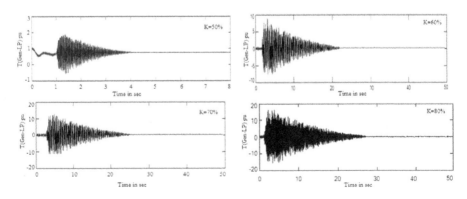

FIGURE 9.12 *T*(Gen-LP)pu during transient state at 50%, 60%, 70%, and 80% series compensation with PI-controlled TCSC.

4 seconds. For maximum value of series compensation (80% of K), the value of generator-LP torque reaches 17 pu and settles at 27 seconds.

At 60% and 70% compensation, the magnitudes of oscillatory torques are 9.5 and 12 pu, and it takes 22 and 25 seconds, respectively, to get completely damped.

CASE II Rotor Speed Deviation ($\Delta\Omega$) during Transient State with PI-Controlled SSSC and TCSC in an IEEE SBM System

Figure 9.13 shows the influence of proposed PI controller in SSSC in damping of rotor speed deviation. Here can be seen a proportionality between the series compensation percentage (K) and settling time. With increase in series compensation level, the rotor speed deviation in pu also increases, thereby increasing its damping time.

The impact of PI-controlled TCSC in damping of rotor speed deviation when the system is subjected to different levels of series compensation is shown in Figure 9.14. For minimum series compensation value of 50%, the rotor speed deviation ($\Delta\omega$) takes 4.2 seconds to completely settle down. For $k = 60\%$, the value of $\Delta\omega$ is 0.030 pu, and it takes 17 seconds to get completely damped. In case of 70% and 80% value of K, the settling time of $\Delta\omega$ oscillations are 20 and 24 seconds, respectively.

Based on the simulation results shown in Figures 9.7 and 9.8 for SSSC and TCSC with conventional PI controller during steady state, it is inferred that SSSC with PI controller damps generator and low-pressure turbine torsional oscillation at 17 and 20 seconds, respectively, at 70% and 80% of K. TCSC with PI controller damps the oscillations at 19 seconds for 70% series compensation, and at 80% series compensation, the damping time of TCSC is 24 seconds.

During transient state with SSSC controller at 60%, 70%, and 80% of K, the rotor speed deviation gets damped at 17, 18, and 22, seconds, respectively, as shown in Figure 9.9. With TCSC controller in an IEEE SBM system, the damping times are 18, 20, and 24 seconds, respectively, at 60%, 70%, and 80% of K. To further test

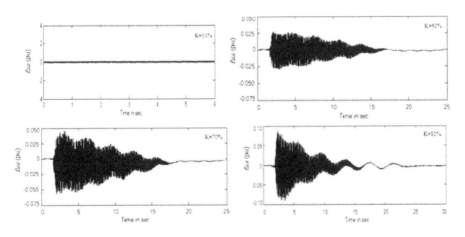

FIGURE 9.13   Rotor speed deviation ($\Delta\omega$) during transient state at 50%, 60%, 70%, and 80% series compensation with PI-controlled SSSC.

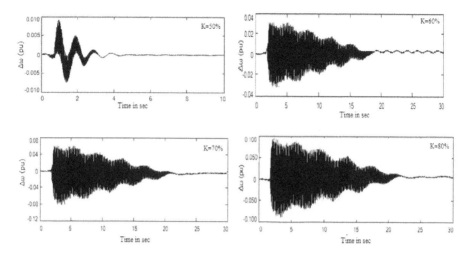

FIGURE 9.14 Rotor speed deviation ($\Delta\omega$) during transient state at 50%, 60%, 70%, and 80% series compensation with PI-controlled TCSC.

the performance of the proposed controller under different loaded condition (25% reduced load), the IEEE SBM system is subjected to transient-state analysis under 70% and 80% values of series compensation.

The impact of PI-controlled TCSC in damping torque between generator and low-pressure turbine in an IEEE SBM system at different levels of series compensation and load is shown in Figure 9.15. For series compensation value of 70%, the magnitude of T(Gen-LP) reaches a magnitude of 7 pu and it takes 14 seconds to completely settle down. For $k = 80\%$, the magnitude of T(Gen-LP) is 12 pu and it takes 22 seconds to get completely damped.

In case of a PI-controlled SSSC, for 70% and 80% values of K, the settling time for T(Gen-LP) magnitude are 8 and 12 seconds, respectively, which is shown in Figure 9.16.

Based on the results with PI controller, it has the drawback of large settling time of the SSR oscillation, which may deteriorate the complete system. To further enhance the performance of SSSC and TCSC in damping SSR oscillations, the PI controller can be replaced with a FLC to find their robustness under different operating environment.

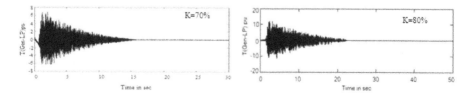

FIGURE 9.15 T(Gen-LP)pu during transient state at 70% and 80% series compensation with PI-controlled TCSC.

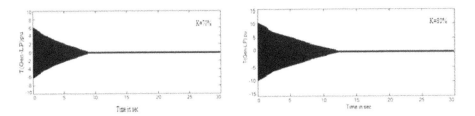

FIGURE 9.16 T(Gen-LP)pu during transient state at 70% and 80% series compensation with PI-controlled SSSC.

9.7 FUZZY LOGIC CONTROLLED SSSC FOR DAMPING SSR OSCILLATIONS

Figure 9.17 shows the schematic diagram of the FLC that consists of the following blocks:

1. Fuzzification
2. Knowledge base
3. Inference engine
4. Defuzzification

9.7.1 FUZZIFICATION MODULE

The fuzzification block changes the input crisp values into fuzzy values. A fuzzy variable uses values that are defined in linguistic manner such as low, medium, high, big, and slow, where each one is confined by a steadily changing membership function. Generally, four basic membership functions are most widely used for fuzzification: triangular, trapezoidal, Gaussian, and generalized bell. In this proposed method, triangular shape is used because it is simple to implement and give good results. The rule base decides the control strategy of the control system, which is

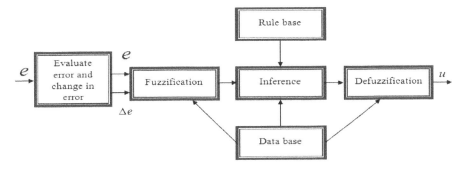

FIGURE 9.17 Structure of a FLC.

usually acquired from expertise knowledge. It comprises an accumulation of fuzzy statements given in a set of if-then rules:

$$R(i): \text{If } a_1 \text{ is } F_1 \text{ and } a_2 \text{ is } F_2 \ldots \text{and } a_n \text{ is } F_n, \text{ then } Y \text{ is } G^{(i)}, i = 1, 2, \ldots, M$$

where $(a_1, a_2, \ldots, a_n)$ are input variables vector, Y is the control variable, M is the number of rules, n is the number of fuzzy variables, and $(F_1, F_2, \ldots, F_n)$ are the fuzzy sets. The fuzzy controller defines the rule base to be fired based on the input condition, and it also computes the output fuzzy variable from the specific input signal. The procedure of converting fuzzy values into crisp values is called defuzzification.

9.7.2 KNOWLEDGE BASE AND INFERENCE ENGINE

From the state variables, error (e), change in error (Δe), and the plant control (u) are inferred. The controllable rules are planned to allocate a fuzzy set to the control output u for every combination of fuzzy sets (e) and (Δe). Table 9.1 shows the rules base, where the rows indicate the rate of change of error and the columns indicate the error. Each pair defines the output value corresponding to du. In Table 4.1, the abbreviations used are as follows: nb– negative big, nm– negative medium, ns– negative small, zr– zero, ps– positive small, pm– positive medium, pb– positive big, b– big, and s– small.

The rule base is the most vital factor of a fuzzy controller. In the inference engine, the fuzzy control strategy is recognized; the rules are framed based on human experience and system performance. For example, when e corresponds to pb membership function and Δe corresponds to pb membership function, it means that the output is more than 1. The error and its change are considered based on the parameter chosen. In this chapter, rotor speed deviation and its derivative are taken as the input signal for the FLC used in SSSC, and variation of line current and its derivative are taken as the input signal in case of TCSC. The above-mentioned rules are written in the

TABLE 9.1
Rule Base of Fuzzy Logic Control

		Error Input e						
du		nb	nm	ns	zr	ps	pm	pb
Change in	nb	b	B	B	B	b	b	b
error	nm	s	B	B	B	b	b	s
input Δe	ns	s	S	B	b	b	s	s
	zr	s	S	S	b	s	s	s
	ps	s	S	B	b	b	s	s
	pm	s	B	B	b	b	b	s
	pb	b	B	B	b	b	b	b

form of a file (.fis) and then called upon in the developed MATLAB/SIMULINK blockdiagram.

9.7.3 Defuzzification

In this part of fuzzy controller, the crisp numeric value, which was used as power system control signal, is created as output of the fuzzy rules. The output of the rule base so obtained is aggregated and defuzzified.Defuzzification is the process of changing a fuzzy set output into a singular value usable output. There are four methods used for defuzzification.

1. Max membership method
2. Centroid method
3. Weighted average method
4. Mean-max membership method

Of these four methods, centroid method is taken here for defuzzification.

The control circuit of fuzzy logic controlled SSSC is shown in Figure 9.18. The rotor speed deviation (Δw) and its derivative arising due to SSR oscillation are taken as input signal; therefore, if speed deviation is controlled, then in parallel, resonance in the subjected system is also compensated. The output of the fuzzy controller in SSSC is to finally vary the firing angle to the SSSC so that the reactance of the line is varied. The line current (I) from the system is connected to PLL, and phase angle

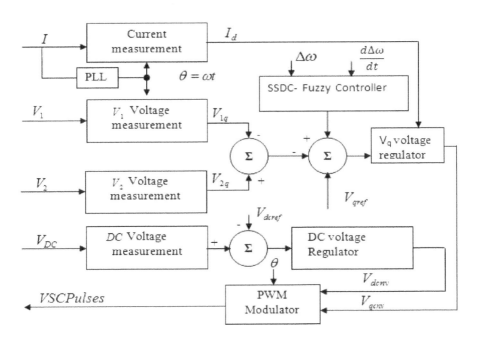

FIGURE 9.18   Block diagram of proposed SSSC controller.

of the transmission line current (θ) is considered from the PLL. The line currents are converted to the direct axis currents and quadrature axis currents, namely, I_d, I_q, with V_q and V_d calculated from Equations (9.3) and (9.4).

The FLC is implemented in the circuit to control the DC link voltage, as given in Equations (9.3) and (9.4).

$$V_a = V_m * \sin \omega t \tag{9.1}$$

$$V_b = V_m * \sin(\omega t + 120) \tag{9.2}$$

$$V_c = V_m * \sin(\omega t - 120) \tag{9.3}$$

The parks transformation can be represented as follows:

$$V_d = \frac{2}{3} * [V_a * \sin(\omega t) + V_b * \sin(\omega t - 120) + V_c * \sin(\omega t + 120)] \tag{9.4}$$

$$V_q = \frac{2}{3} * [V_a * \cos(\omega t) + V_b * \cos(\omega t - 120) + V_c * \cos(\omega t + 120)] \tag{9.5}$$

The rotor speed deviation ($\Delta \omega$) and its derivative ($d\Delta \omega/dt$) are taken as the input to the fuzzy controller. The DC voltage across the series converter capacitor V_{DC} is compared with the actual V_{dcref} and the error is fed to DC regulator. V_{dcon} thus obtained from the DC regulator is converted into three-phase voltage using inverse Park's transformation. In PWM block operation, the necessary pulses required to control the SSSC are generated in MATLAB/SIMULINK software package.

9.8 FUZZY LOGIC CONTROLLED TCSC FOR DAMPING SSR OSCILLATIONS

A TCSC with a FLC is implemented to select the value of firing angle (α) to damp the SSR oscillation. The schematic diagram of current controller with FLC is shown in Figure 9.19.The TCSC current control TCSC moduleutilizes a controllable thyristor-controlled reactor (TCR) branch in parallel with series capacitor for enabling fast and effective variation of the device reactance. The value of firing angles of the thyristors will decide the mode of operation like inductive and capacitive control modes. The ratio between the effective fundamental reactance X_{eff} of the device

and the reactance X_c of series capacitor is called boost factor (K_b) of the device, described by the relation given in Equation (9.6):

$$K_b = \frac{X_{eff}}{X_c} \tag{9.6}$$

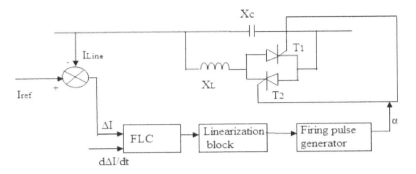

FIGURE 9.19   Block diagram of TCSC with current controller.

Therefore, if the firing angles of the thyristors are varied, it will vary $t\,X_{eff}$, which will result in the modification of transmission line impedance for controlling power flow, for improving system stability, and, in my consideration, for damping SSR oscillations.

TCSC device operating in constant current control model was designed to substitute for a portion of the fixed series compensation in the IEEE SBM system. The change in line current and its derivative is taken as the input to the fuzzy controller. The inputs to the fuzzy controller are change in line current and its derivative.

9.9   CASE STUDY

For validation of robustness of fuzzy logic controlled SSSC and TCSC device in improving SSR oscillations damping, IEEE SBM systems shown in Figures 9.18 and 9.19 are considered. The time domain simulation analysis using MATLAB software is done under steady-state and transient conditions. The occurrence of torsional interaction that was observed from eigenvalue analysis in an IEEE SBM system with different levels of series compensation is validated with time domain simulation analysis. For evaluating the proposed controller's performance, the system response is compared under different stateswith FLC-based SSSC and TCSC.

9.9.1   STEADY-STATE SSR ANALYSIS IN STEAM TURBINE
GENERATOR WITHOUT CONTROLLERS

The time domain simulations are performed for 50%, 60%, 70%, and 80% series compensation levels K. Based on Figure 9.20, the torque between generator and low-pressure turbine[T(Gen-LP)] pu shows the existence of SSR oscillation at values of 80% and 70% series compensation (K), respectively. It can be observed that there is no oscillation occurring at 50% and 60% values of K.

From Figure 9.21, the rotor speed deviation ($\Delta\omega$) tends to originate at 70% and 80% values of K, and reaches the values of 0.16 and 0.20 pu, respectively, at 40 seconds each and proves the existence of SSR oscillations and its subsequent effects.

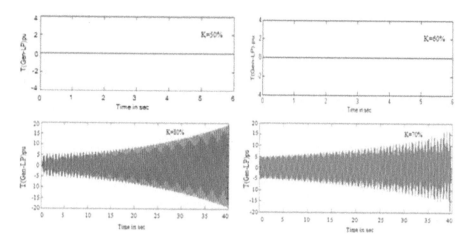

FIGURE 9.20 T(Gen-LP)pu during steady state at 50%, 60%, 70%, and 80% series compensation without controllers.

9.9.2   Steady-State SSR Analysis in Steam Turbine Generator with Controllers

Since during the steady-state conditions, the existence of SSR oscillations is observed at 70% and 80% series compensation levels K, the same system is now tested with fuzzy logic controlled SSSC and TCSC to analyze their effectiveness in damping SSR-related impact.

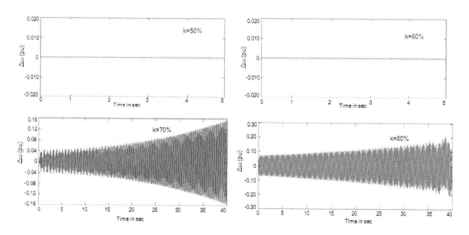

FIGURE 9.21   Rotor speed deviation ($\Delta\omega$) during steady state at 50%, 60%, 70%, and 80% series compensation without controllers.

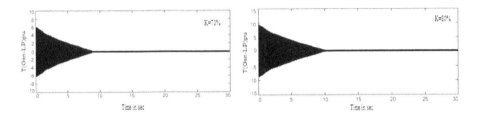

FIGURE 9.22 *T*(Gen-LP)pu during steady state at 70% and 80% series compensation with fuzzy logic controlled SSSC.

CASE I Torque between Generator and Low-Pressure Turbine *T*(Gen-LP)pu during Steady State with Fuzzy-Controlled SSSC and TCSC in an IEEE SBM System

As observed from Figure 9.22, with SSSC controller, the Gen-LP torque oscillation at $K = 70\%$ attains a maximum value of 6 pu, and due to the effect of controller, the oscillation gets damped at 8 seconds. At $K = 80\%$, Gen-LP torque oscillation takes 11 seconds to get completely damped from a magnitude of 9 pu.

The observation from Figure 9.23 shows the Gen-LP torque oscillation at $K = 70\%$ reaches a maximum value of 7 pu, and due to the effect of TCSC controller, the oscillation gets completely damped at 13 seconds. At $K = 80\%$, Gen-LP torque oscillation takes 12 seconds to get completely damped from a magnitude of 10 pu.

CASE II Rotor Speed Deviation (ΔΩ) during Steady State with Fuzzy-Controlled SSSC and TCSC in an IEEE SBM System

The rotor speed deviation ($\Delta\omega$) observed from Figure 9.24 tends to reach a maximum value of 0.015 pu at $K = 70\%$ and 0.032 pu at $K = 80\%$, and due to the impact of SSSC controller, the oscillation gets damped at 6 and 9 seconds, respectively.

According to Figure 9.25, rotor speed deviation ($\Delta\omega$) reaches a maximum value of 0.040 pu at $K = 70\%$ and 0.055 pu at $K = 80\%$. Due to the impact of TCSC controller, the oscillation gets damped at 12 and 20 seconds, respectively.

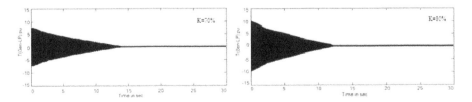

FIGURE 9.23 *T*(Gen-LP)pu during steady state at 70% and 80% series compensation with fuzzy logic controlled TCSC.

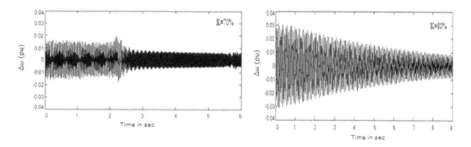

FIGURE 9.24 Rotor speed deviation ($\Delta\omega$) during steady state at 70% and 80% series compensation with fuzzy logic controlled SSSC.

9.9.3 TRANSIENT-STATE SSR EFFECT ANALYSIS IN STEAM TURBINE GENERATOR WITHOUT CONTROLLERS

A three-phase fault of 0.05 seconds duration is simulated at the series-compensated line of the study system. Figure 9.26 shows the impact of the fault on the Gen-LP torque oscillation at different levels of K. The value of K is set to 50%, 60%, 70%, and 80%. As K increased from 50% to 80%, the oscillation at the Gen-LP shaft section also increases. At 50% series compensation, it is observed that the Gen-LP torsional oscillation reaches a maximum value of 1.2 pu and gets damped at 4 seconds, and for all other remaining values of K, the oscillations get increased, as depicted in Figure 9.26.

A similar fault study is initiated on the IEEE SBM system and the results are analyzed. Figure 9.27 shows the simulation results of the impact of fault (three-phase to ground fault) on the rotor speed deviation ($\Delta\omega$) at different levels of series compensation K. The value of K is set to 50%, 60%, 70%, and 80%. At 50% and 60% series compensation levels, ($\Delta\omega$) reaches a maximum value of 0.009 and 0.018 pu, respectively, and reaches steady state from there on. For remaining values of K at 70% and 80%, the rotor speed deviation gets increased, as depicted in Figure 9.27.

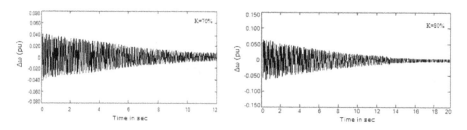

FIGURE 9.25 Rotor speed deviation ($\Delta\omega$) during steady state at 70% and 80% series compensation with fuzzy logic controlled TCSC.

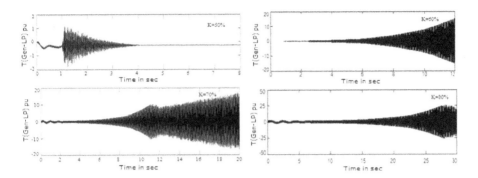

FIGURE 9.26 T(Gen-LP)pu during transient state at 50%, 60%, 70%, and 80% series compensation without controller.

9.9.4 TRANSIENT-STATE SSR ANALYSIS IN STEAM TURBINE GENERATOR WITH CONTROLLERS

In order to demonstrate the influence of fuzzy logic controlled SSSC and TCSC in damping SSR oscillation effects during transient state, simulation runs were done with different levels of series compensation percent K.

CASE I T(GEN-LP) PU DURING TRANSIENT STATE WITH FUZZY-CONTROLLED SSSC AND TCSC IN AN IEEE SBM SYSTEM

The plots for T(Gen-LP) with different values of K with respect to settling time are shown in Figure 9.28. All the plots shown in Figure 9.28 are simulated with MATLAB software for an IEEE SBM system with fuzzy logic controlled SSSC.

Based on Figure 9.28, due to the influence of proposed controllers, the damping of turbine-generator oscillations is achieved at all levels of K. But one important

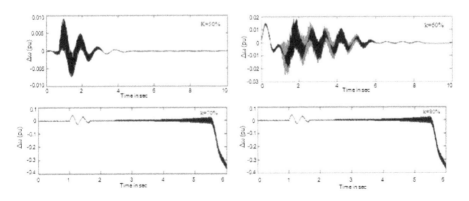

FIGURE 9.27 Rotor speed deviation ($\Delta\omega$) during transient state at 50%, 60%, 70%, and 80% series compensation without controller.

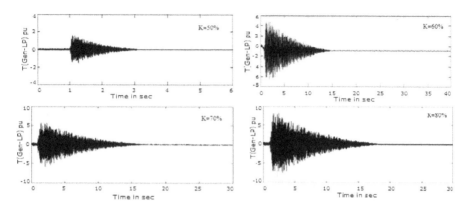

FIGURE 9.28 *T*(Gen-LP)pu during transient state at 50%, 60%, 70%, and 80% of *K* with fuzzy logic controlled SSSC.

observation made is that as *K*-value increases from 50% to 80%, turbine-generator oscillatory torque magnitude also increases, thereby increasing the settling time of oscillatory torque.

The effect of fuzzy-controlled TCSC in damping of torque between generator and low-pressure turbine is shown in Figure 9.29. At 50% of series compensation, it could be observed that the value of generator-LP torque reaches a maximum of 1.5 pu and gets damped completely at 3.8 seconds. For maximum value of series compensation *K* = 80%, the value of Generator-LP torque reaches 11 pu and settles at 22 seconds. At 60% and 70% compensation, the magnitudes of oscillatory torques are 7 and 10 pu, respectively, and takes 15 and 22 seconds, respectively, to get completely damped.

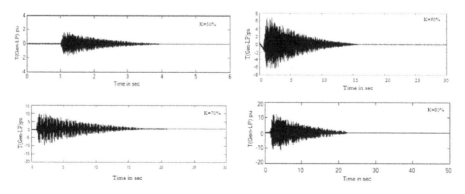

FIGURE 9.29 *T*(Gen-LP)pu during transient state at 50%, 60%, 70%, and 80% of *K* with fuzzy logic controlled TCSC.

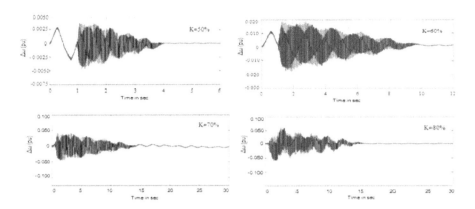

FIGURE 9.30   Rotor speed deviation ($\Delta\omega$) during transient state at 50%, 60%, 70%, and 80% of K with fuzzy-controlled SSSC.

CASE II ROTOR SPEED DEVIATION ($\Delta\Omega$) DURING TRANSIENT STATE WITH FUZZY-CONTROLLED SSSC AND TCSC IN AN IEEE SBM SYSTEM

Figure 9.30 shows the influence of proposed fuzzy controller in SSSC in damping of rotor speed deviation. Here can be seen a proportionality between the series compensation percentage (K) and settling time. With increase in series compensation level, the rotor speed deviation in pu increases, thereby increasing its damping time.

The impact of fuzzy-controlled TCSC in damping of rotor speed deviation when the system is subjected to different levels of series compensation is shown in Figure 9.31. For minimum series compensation value of 50%, the rotor speed deviation takes 4.2 seconds to completely settle down. For $K = 60\%$, the value of rotor speed deviation is 0.025 pu and takes 16 seconds to get completely damped. In case of

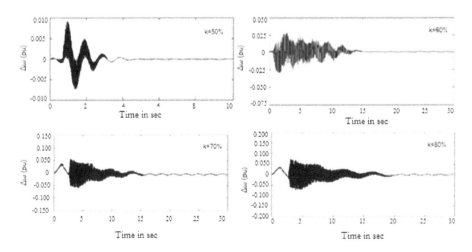

FIGURE 9.31   Rotor speed deviation ($\Delta\omega$) during transient state at 50%, 60%, 70%, and 80% of K with fuzzy-controlled TCSC.

TABLE 9.2

T(Gen-LP)pu during Steady State at Different Levels of Series Compensation (*K*) with Controller (Fuzzy-Controlled SSSC and TCSC) and without Controller

	Without Controller		With SSSC		With TCSC	
Results *K*	Magnitude (pu)	Time (s)	Magnitude (pu)	Time (s)	Magnitude (pu)	Time (s)
50	0	0	–	–	–	–
60	0	0	–	–	–	–
70	16.0	40.0	6.0	8.0	7.0	13.0
80	20.0	40.0	9.0	10.0	10.0	12.0

70% and 80% values of *K*, the settling time of rotor speed oscillations are 16 and 21 seconds, respectively. Tables 9.2–9.5 shows the settling time of torsional oscillations as well as the rotor speed deviation in an IEEE SBM system with and without TCSC and SSSC based FLC.

For minimum series compensation value of 50%, the rotor speed deviation takes 4.2 seconds to completely settle down. For *K* = 60%, the value of rotor speed deviation is 0.025 pu and it takes 16 seconds to get completely damped. In case of 70% and 80% values of *K*, the settling time of rotor speed oscillations are 16 and 21 seconds, respectively.

To further test the performance of the proposed controller under different loaded conditions, the IEEE SBM system is subjected to transient-state analysis under 70% and 80% values of series compensation.

The impact of fuzzy-controlled TCSC in damping rotor speed deviation when the IEEE SBM system is subjected to different levels of series compensation is shown in Figure 9.32. For series compensation value of 70%, the rotor speed deviation takes

TABLE 9.3

Rotor Speed Deviation (Δω)pu during Steady State at Different Levels of Series Compensation (*K*) with Controller (Fuzzy-Controlled SSSC and TCSC) and without Controller

	Without Controller		With SSSC		With TCSC	
Results *K*	Magnitude (pu)	Time (s)	Magnitude (pu)	Time (s)	Magnitude (pu)	Time (s)
50	0	0	0	0	0	0
60	0	0	0	0	0	0
70	0.16	40.0	0.015	6.0	0.040	12.0
80	0.20	40.0	0.032	9.0	0.055	20.0

TABLE 9.4

T(Gen-LP)pu during Transient State at Different Levels of Series Compensation (K) with Controller (Fuzzy-Controlled SSSC and TCSC) and without Controller

Results K	Without Controller		With SSSC		With TCSC	
	Magnitude (pu)	Time (s)	Magnitude (pu)	Time (s)	Magnitude (pu)	Time (s)
50	1.2	4.0	1.5	3.0	1.5	3.80
60	15.0	12.0	5.0	14.0	7.0	15.0
70	18.0	20.0	6.0	16.0	10.0	22.0
80	25.0	30.0	8.0	18.0	11.0	22.0

16 seconds to completely settle down from a magnitude of 0.040 pu. For $K = 80\%$, the magnitude of rotor speed deviation is 0.049 pu and it takes 16 seconds to get completely damped.

In case of a fuzzy-controlled SSSC, for 70% and 80% values of K, the settling time of rotor speed oscillations are 9 and 12 seconds, respectively, as shown in Figure 9.33.

The impact of fuzzy-controlled TCSC in damping torque between generator and low-pressure turbine in an IEEE SBM system at different levels of series compensation and load is shown in Figure 9.34. For series compensation value of 70%, the magnitude of T(Gen-LP) takes 17 seconds to completely settle down from a magnitude of 8 pu. For $K = 80\%$, the magnitude of T(Gen-LP) is 10 pu and it takes 20 seconds to get completely damped.

In case of a fuzzy-controlled SSSC, for 70% and 80% values of K, the settling time for T(Gen-LP) magnitude are 11 and 13 seconds, respectively, which is shown in Figure 9.35.

TABLE 9.5

Rotor Speed Deviation ($\Delta\omega$)pu during Transient State at Different Levels of Series Compensation (K) with Controller (Fuzzy-Controlled SSSC and TCSC) and without Controller

Results K	Without Controller		With SSSC		With TCSC	
	Magnitude (pu)	Time (s)	Magnitude (pu)	Time (s)	Magnitude (pu)	Time (s)
50	0.009	4.5	0.0027	4.0	0.009	4.20
60	0.018	6.0	0.020	10.0	0.025	16.0
70	−0.400	6.0	0.045	15.0	0.055	16.0
80	−0.400	6.0	0.060	16.0	0.075	21.0

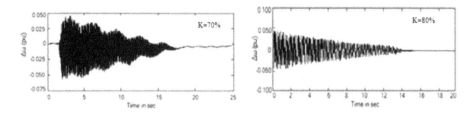

FIGURE 9.32 Rotor speed deviation (Δω) during transient state at 70% and 80% series compensation with fuzzy logic controlled TCSC.

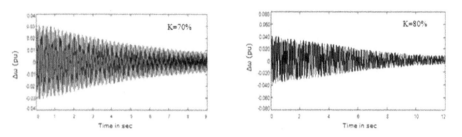

FIGURE 9.33 Rotor speed deviation (Δω) during transient state at 70% and 80% series compensation with fuzzy logic controlled SSSC.

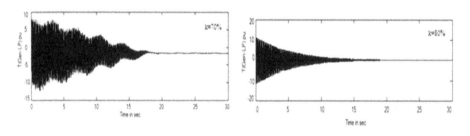

FIGURE 9.34 T(Gen-LP)pu during transient state at 70% and 80% series compensation with fuzzy logic controlled TCSC.

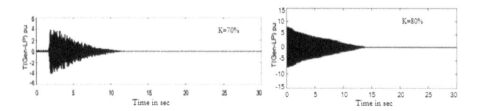

FIGURE 9.35 T(Gen-LP)pu during transient state at 70% and 80% series compensation with fuzzy logic controlled SSSC.

9.10   CONCLUSION

This chapter explains the robustness of SSSC and TCSC in the IEEE SBM system, under different operating environment. The effects of PI-controlled SSSC and TCSC are also discussed with case studies. Various data observed from simulations are tabulated to find the individual effectiveness of SSSC and TCSC. It is inferred from the tabulated data that the FLC-based SSSC can contribute to improved damping characteristics in terms of reduced settling time over a wide range of operating conditions as compared to fuzzy logic controlled TCSC.

9.11   SUMMARY

i. A conventional PI controller in damping SSR-related oscillation (torsional oscillation between generator and low-pressure turbine) and rotor speed deviation in an IEEE SBM system are analyzed.According to theresults, it can be concluded that a PI-controlled SSSC is superior to a PI-controlled TCSC in damping SSR-related oscillation.

ii. The influence of fuzzy logic controlled TCSC and SSSC in damping SSR oscillations at generator-low pressure turbine oscillation and rotor speed deviation in an IEEE SBM system is analyzed.

iii. Further analysis is also performed with different loads in an IEEE SBM system.It is concluded that FLC-based SSSC has superior oscillations damping capability compared to FLC-based TCSC.

REFERENCES

Amini, Z & Kargar, A, 2013, 'Reduction of sub-synchronous resonance using artificial neural network', International Journal of Multidisciplinary Sciences and Engineering, vol. 4, no. 10, pp. 6–9.

Hosseini, SMH, Samadzadeh, H, Olamaei, J & Farasadi, M, 2013, 'SSR mitigation with SSSC thanks to fuzzy control', Turkish Journal of Electrical Engineering & Computer Sciences, vol. 21, pp. 2294–2306.

Kumar, S, Kumar, N & Jain, V, 2011, 'Comparison of various auxiliary signals for damping subsynchronous oscillations using TCR-FC', Energy Procedia, vol. 14, pp. 695–701.

Li, J & Zhang, X, 2016, 'Impact of increased wind power generation on subsynchronous resonance of turbine-generator units', Journal of Modern Power Systems and Clean Energy, vol. 4, pp. 219–228.

Sindhu Thampatty, KC, Nandakumar, MP & Cheriyan, EP, 2011, 'Adaptive RTRL based neurocontroller for damping subsynchronous oscillations using TCSC', Engineering Application of Artificial Intelligence, vol. 24, pp. 60–76.

Varma, RK & Auddy, S, 2006a, 'Mitigation of subsynchronous resonance using PMU-acquired remote generator speed', Power India Conference, IEEE, pp. 1–8.

Varma, RK & Auddy, S, 2006b, 'Mitigation of subsynchronous oscillations in a series compensated wind farm with static var compensator', Proceedings of IEEE Power Engineering Society General Meeting, IEEE, pp. 1–7.

Vivek, S & Selve, V, 2014, 'SSR mitigation and damping power system oscillation in a series compensated wind generation system', 2014 IEEE National Conference on Emerging Trends in New & Renewable Energy Sources and Energy Management (NCET NRES EM), IEEE, Chennai, pp. 32–38.

Zhu, X, Jin, M, Kong, X et al., 2018, 'Subsynchronous resonance and its mitigation for power system with unified power flow controller', Journal of Modern Power Systems and Clean Energy, vol. 6, pp.181–189.

10 Fuzzy Logic Controlled TCSC and SSSC in Wind Power Integrated IEEE SBM System

LEARNING OUTCOME

i. To learn about wind energy conversion systems (WECS).
ii. To study about wind power integrated IEEE SBM system under steady-state and transient-state conditions.
iii. To have knowledge on fuzzy logic controlled SSSC and TCSC.

10.1 INTRODUCTION

The introduction of distributed generation has consequences not only on the distribution network but also on the transmission grid as well as on the rest of the generators. Whenever the wind generators are installed in the existing power system, power system engineers have to analyze the worst operating scenarios in order to prevent the power system from adverse effect. Cheng et al (2014) proposed a detailed review about Wind energy conversion systems. With the objective of analyzing such scenarios, the modified IEEE Second Benchmark Model (SBM) system is developed. There are two types of induction generators used in wind farms: single-cage and double-cage. Double-fed induction generator (DFIG) is integrated into the existing developed IEEE SBM system that consists of a two-stage four-mass steam turbine generator system feeding a series-compensated parallel transmission line. The test system so obtained after integration is termed as modified IEEE Second Benchmark Model system.

The various research findings discussed in literature survey have conveyed at the outset about the system vulnerability toward subsynchronous resonance (SSR) oscillations. With the integration of wind energy system into an existing system having conventional turbine generator system, the level of series compensation will have SSR impact on either turbine generator system or wind turbine generators.

Therefore, the need for analyzing the possibility of SSR at various compensation levels before planning the interconnection of wind farms to already existing system arises. It is therefore of paramount importance to see the effect of SSR on wind and conventional turbine generator systems and its response toward various mitigation methods. Flexible alternating current transmission system (FACTS) devices, namely, SSSC and thyristor-controlled series capacitor (TCSC), have been used for controlled series compensation along with controllers to damp or mitigate SSR oscillations at a faster rate. The analysis is carried out with an IEEE SSR SBM system model along with low-power DFIG wind farm using the MATLAB software package. The aim is to clarify the fact that without SSSC and TCSC and its proposed controller, the rotor of the four-mass steam turbine systems will suffer prolonged oscillation and the fluctuation in the system also increases. Also, the negligible impact of SSR-related oscillations in low-power wind generation unit is to be proved here. To simulate and find the effectiveness of individual FACTS device under transient conditions, a three-phase ground fault is made to occur in a transmission line at 1 second and removed after 5 milliseconds using MATLAB/SIMULINK software.

Jang et al. (1997) work on Neuro-Fuzzy and Soft Computing serves as a guide to researchers. Rahim et al. (2004) proposed a topology for reduction of subsynchronous resonance. Gao et al (2008) proposed an energy management strategy that is also based on fuzzy logic.

Wang and Truong (2013) presented his work to improve the stability of a power system based on an adaptive-network-based fuzzy inference system. Gahramani et al. (2013) proposed a mitigation methodology of SSR and LFO which is also based on fuzzy logic.

Pachauri et al. (2016) proposed a control for wind turbine system using PI/fuzzy control. Chilwal et al. (2020) presented a survey of fuzzy logic inference system and other computing techniques for agricultural diseases. This chapter explains about the performance of SSSC and TCSC with fuzzy logic controller(FLC) in damping SSR oscillations in the modified IEEE SBM system. The earlier studies prove that (Fan and Miao, 2012) series-compensated transmission lines connected to induction generator (IG) based wind farms are subjected to SSR oscillations, depending upon the level of series compensation and power generation capacity of the wind farm. The time domain simulation analysis is performed on the modified system, with fuzzy logic controlled SSSC and TCSC. In both cases, it is possible to damp SSR oscillations created by series capacitors with the help of an auxiliary damping controller (FLC) in TCSC and SSSC. Unlike the old traditional hard computing techniques, the soft computing techniques used have the advantage of nonrequirement of a mathematical model and hence it is becoming more familiar as system identification methodology. Fuzzy logic based soft computing approach in FACTS device is analyzed in this chapter for system subsynchronous oscillation identification and mitigation.

The SSR damping and mitigation of the power system is improved by injecting the additional voltage and changing the reactance of the line. The amount of voltage and firing angle to be injected can be calculated using fuzzy logic. The MATLAB/SIMULINK software program is used for verifying the effectiveness of control methodology. The simulation results proved the control ability of TCSC and SSSC with FLC and its robustness to fault and change in operating situations.

10.2 OVERVIEW OF WIND ENERGY CONVERSION SYSTEMS (WECS)

The following are the three types of generators that are commonly employed in commercial wind turbines:

1. Fixed-speed wind energy conversion systems (FSWECS) or type-1 wind turbines
2. Partially variable-speed wind energy conversion systems (PVWECS) or type-2 wind turbines
3. Variable-speed wind energy conversion systems (VSWECS) based on DFIG (type-3 wind turbines) and permanent magnet synchronous generator (PMSG) (type-4 wind turbines)

10.2.1 TYPE-1 WIND TURBINE

Type-1 wind turbine or a fixed-speed wind turbine contains a rotor and a squirrel-cage induction generator (SCIG), connected through a gearbox, as shown in Figure 10.1. The stator winding of the generator is connected to grid, and the generator slip varies with respect to the generated power. The speed variations of this type of generator are very small (just 1–2%).

The squirrel-cage generator extracts reactive power from the grid, which is not desirable, in the case of weak networks. Hence, for compensation of this reactive power consumption, capacitor banks are installed along with squirrel-cage generators, as shown in Figure 10.1. To reduce the transient current during energization of the machine, a set of antiparallel thyristors are used to build up the magnetic flux slowly. The advantages of fixed-speed WECS are that it is simple in operation, it is robust, and it provides reliable operation. However, owing to its fixed-speed

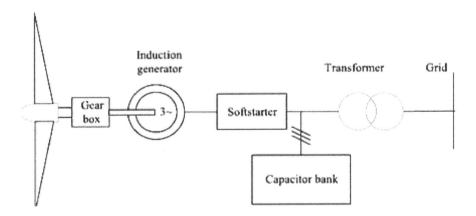

FIGURE 10.1 Structure of type-1 wind turbine.

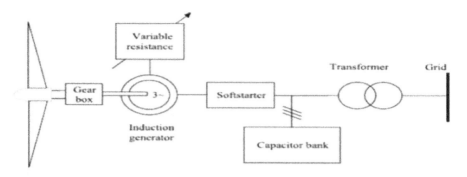

FIGURE 10.2   Structure of type-2 wind turbines.

operation, its major drawback is that it leads to electrical fluctuations into the grid in case of variation in wind speed.

10.2.2   TYPE-2 WIND TURBINE

Type-2 wind turbines are variable-speed WECS, which are fitted with a wound-rotor induction generator (WRIG) and variable external rotor resistance, as displayed in Figure 10.2. The distinct feature of this type of WECS is its additional variable controllable resistance. Having this variable feature, the total resistance that is the summation of internal and external resistance is changeable; hence, the slip of the machine is easily controllable. This unique feature makes this turbine more reliable in controlling its mechanical characteristics. Depending on the range of additional resistance inserted, the speed of the machine can be increased 10% above its rated speed.

10.2.3   TYPE-3 WIND TURBINE

A DFIG wind turbine configuration is shown in Figure 10.3, which has a WRIG with slip rings for making a path for both-way current flows in the rotor winding. To attain a variable-speed operation, a variable voltage at slip frequency is injected to the rotor via a variable-frequency power converter. The power converter separates the network electrical frequency and rotor mechanical frequency, so as to make wind turbine accessible to operate at various speed. The main advantage of DFIG system is its capability of delivering power to the grid from both stator and rotor when operated in supersynchronous mode.

10.2.4   TYPE-4 WIND TURBINE

The type-4 wind turbine is variable-speed wind turbine with PMSG, as shown in Figure 10.4, which has the rotor and generator shafts coupled directly. The generator is low-speed multi-pole synchronous generator that can be either an electrically excited synchronous generator or a permanent magnet generator. For flexible-speed

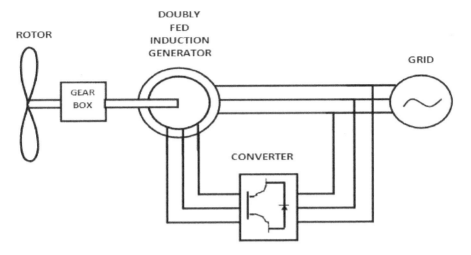

FIGURE 10.3   Structure of type-3 wind turbine.

operation, the synchronous generator is connected to the grid via a flexible-frequency converter. This application of variable-frequency power converter makes the grid frequency to remain unchanged in spite of variation in wind speed.

10.3   DFIG CONTROL STRATEGIES

The notable advantage that can be observed in the design of DFIG in comparison with type-3 and type-4 wind turbines is that it uses a low-cost power electronics converter in the rotor, whereas in the PMSG-based WECS, a larger electronic converter is used and the full 100% of the power generated has to pass through it. However, the main drawback of DFIG systems is their unreliable gearbox system. The usage

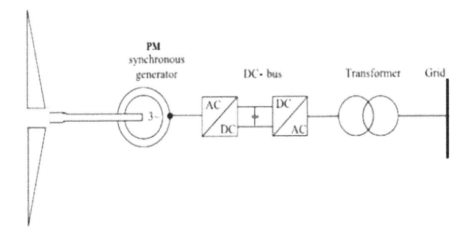

FIGURE 10.4   Structure of type-4 wind turbine.

of DFIG in power generation has recently increased due to its reliable features necessary for grid integration. The majority of the researchers' contribution to the field of DFIG is in DFIG modeling for stability analysis, maximum power point tracking algorithms, unbalanced network of DFIG operation, and less SSR study conducted in DFIG.

According to Figure 10.5, a DFIG consists of back-to-back converter at rotor circuit through a common DC link through power converters and a stator is connected to the grid directly.

The function of electrical control system in the above model is to make sure that desirable voltage and current are maintained at the DC common link and grid side. There are different control strategies available in DFIG, which are discussed here.

10.3.1 POWER CONTROL STRATEGY

In this method, the speed of the wind turbine is controlled by pitch control. The turbine mechanical power output is given in Equation (10.1):

$$W = \frac{1}{2}\rho A V^3 C_P \tag{10.1}$$

where ρ is the air density in (kg/m^3), A is the swept area of wind turbine in m^2, V is the wind speed (m/s), and C_P is the aerodynamic efficiency (power coefficient), which depends on pitch angle (β) and tip speed ratio – TSR (λ).

Figure 10.6 shows the characteristics between power coefficient C_P and the TSR. Based on Figure 10.6, the value of C_P reaches peak for a specific value of TSR. Hence, there is need for changing the turbine rotational speed with respect to wind speed for harvesting maximum mechanical power from the wind. If the speed of wind is below the rated value, the pitch angle is set to low value and the generator speed is electrically controlled either through vector or by direct torque control (DTC) for maximum power

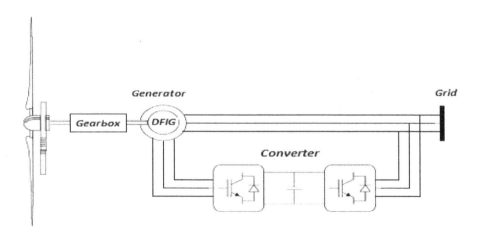

FIGURE 10.5   Model of DFIG wind power plant.

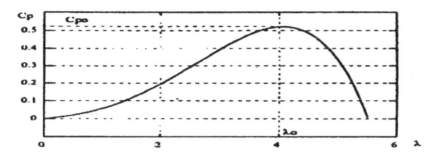

FIGURE 10.6 Typical power coefficient versus TSR curve.

extraction. At high wind speeds, pitch control is enabled for limiting the power output within rated value to protect the generator from mechanical damage.

10.3.2 Vector Control Strategies of DFIG

It is based on a transformation of three-phase variables into dq reference frame. The reference frame is aligned to the stator or rotor flux of the machine rotating at synchronous speed. The vector control has the feasibility of independent active and reactive power flow control between the grid and generator.

10.3.3 Direct Torque Control (DTC) of DFIG

This approach requires the information regarding stator flux and torque for machine control. The stator flux is obtained by integrating the stator voltage. The main disadvantage of DTC method is its inferior performance during initial starting and at very low speed. There are several approaches like usage of dither signal, predictive techniques, or switching table that can be used to regularize the problems faced in DFIG.

10.4 DESCRIPTION OF AN IEEE SBM SYSTEM WITH WIND TURBINE GENERATOR

The power system network considered for analysis is a modified IEEE SBM, which has a low-power DFIG wind farm (100 MW) aggregated with a conventional four-mass steam turbine generator unit, as shown in Figure 10.7.

This modified IEEE SBM system shown in Figure 10.7 will be tested under various levels of series compensation during steady-state and transient conditions to identify the existence of SSR-related oscillation and its impact on DFIG and four-mass steam turbine generator unit. For damping the SSR-related oscillations, FLC compensation devices discussed in Chapter 4 will be subjected to test for validating its SSR damping ability in this modified IEEE SBM system.

The modified IEEE SBM system with SSSC and TCSC as shown in Figures 10.8 and 10.9, respectively, will be tested under both steady-state and transient working conditions.

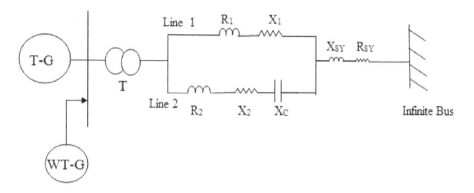

FIGURE 10.7   IEEE SBM model aggregated with wind turbine generator.

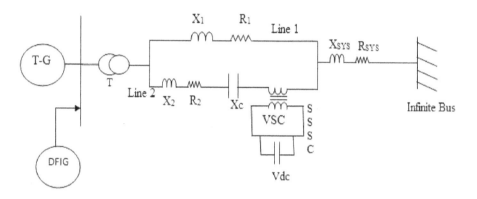

FIGURE 10.8   IEEE SBM model aggregated with wind turbine generator and SSSC.

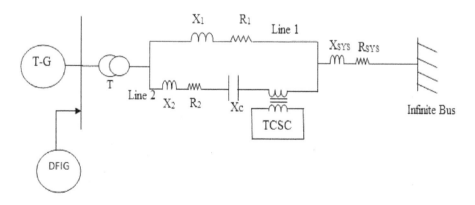

FIGURE 10.9   IEEE SBM model aggregated with wind turbine generator and TCSC.

The modified IEEE SSR SBM system consists of a four-mass two-stage steam turbine generator and a DFIG wind generation unit connected to an infinite bus through two transmission lines, one of which is variable seriescompensated. The mechanical subsystem consists of four-mass two-stage steam turbines. The compensation capacitor (X_c) introduces subsynchronous oscillations, which was discussed in Chapter 2 in detail. Once a three-phase fault has been applied and cleared, there exists a phenomenon exciting the torsional oscillation of the multi-mass shaft.

10.5 FUZZY LOGIC CONTROLLED SSSC AND TCSC IN A MODIFIED IEEE SBM SYSTEM

There are quite a large number of countermeasure techniques originally proposed for the reduction of torsional interactions in large synchronous generators coupled with multistage turbines (Hosseini et al., 2013). The countermeasures can be divided into three groups:

1. Unit tripping countermeasures
2. Non-unit tripping countermeasures
3. Conceptual countermeasures

The application of FACTS controllers for the SSR mitigation comes under the category of conceptual method of SSR mitigation. Two types of FACTS devices, namely, TCSC and SSSC, with FLDC controllers are proposed and tested individually in the modified IEEE system shown in Figure 10.7.SSSC is another most important FACTS controller, which is based on voltage source converter (VSC) as shown in Figure 10.10.

Typically, the series-injected voltage V_i or V_{SSSC} will be quite small compared to the line voltage. Without an extra energy source, SSSC has the ability to inject a variable voltage, 90 degrees in leading or lagging with the current. Due to the ability of SSSC to change its reactance characteristic from capacitive to inductive, it is very effective in controlling power flow in power systems. When an auxiliary stabilizing signal is superimposed on the power flow control function, SSSC will improve power system

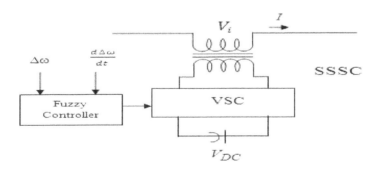

FIGURE 10.10　Basic scheme of SSSC with FLC.

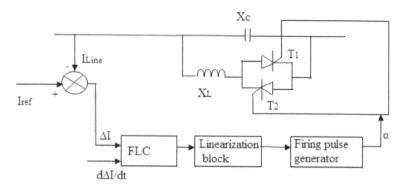

FIGURE 10.11   Basic scheme of TCSC with FLC.

oscillation damping. This control methodology was used in Chapter 9 under Sections 9.3 and 9.4 for testing fuzzy logic controlled SSSC and TCSC for SSR oscillations mitigation in an IEEE SBM system. The identical control strategy is used in this proposed modified IEEE SBM system, as shown in Figures 10.8 and 10.9 with FLC in TCSC and SSSC. TCSC utilizes a controllable TCR with series-connected capacitor to enable fast and flexible variation of effective reactance of the device, as shown in Figure 10.11.

The inductive and capacitive control modes of operation depend upon the firing angles instant of the thyristors. Hence, a proper adjustment of firing angles of the thyristor yields modification of transmission line impedance for controlling power flow and improving system stability, thereby enhancing power transfer capacity and SSR oscillations damping. The control strategies of TCSC with an auxiliary FLDC are same like the one that was implemented in Chapter 9 for SSR oscillation damping. Here "modified IEEE SBM system integrated with a DFIG" is considered.

10.6   CASE STUDY

For validation, the modified IEEE SBM system is developed and SSR analysis was conducted with different levels of series compensation during steady-state and transient-state conditions, using MATLAB software. For analyzing the transient stability performance of SSSC and TCSC with FLC, a three-phase to ground fault is initiated at series-compensated line.

The considered study system is modeled using the inbuilt block sets given in the MATLAB software package. The time domain simulations analysis is carried out to find the potential for the SSR at wind farm as well as at the four-mass steam turbine generator. The simulations are carried out in two parts: (i) steady state and (ii) transient state.

10.6.1   Steady-State SSR Analysis at Wind Turbine Generator in a Modified IEEE SBM System

Here for case study, a 100-MW wind farm is integrated into the existing IEEE SBM system and simulation study is performed to determine the impact of FACTS device controlled series compensation (K) using MATLAB software.

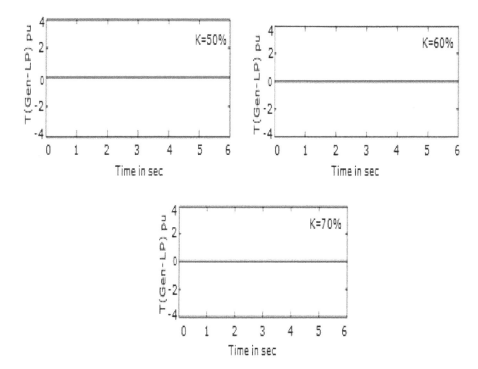

FIGURE 10.12   Electromagnetic torque (T_e) at 100-MW wind turbine generator for $K = 50\%$, 60%, and 70% during steady state.

Figure 10.12 shows the time domain simulation and based on this figure, the electromagnetic torque is stable in steady state with different levels of series compensation. Also, the rotor speed deviation remains to be stable under different levels of K, as depicted in Figure 10.13.

10.6.2   SSR Analysis at Wind Turbine Generator in a Modified IEEE SBM System during Transient State

Again,100-MW wind farm is integrated into the existing IEEE SBM system, as shown in Figure 10.7, and is subjected to simulation, and the impact of SSR oscillations during transient state at various levels of series compensation using MATLAB software was studied. Figure 10.14 shows the time domain simulation, and from this figure it is found that the electromagnetic torque reaches a maximum of 1.8 pu and the oscillations get completely damped at 3.2 seconds even at 70% level of series compensation in a line.

With respect to 50% and 60% levels of series compensation (K), it can be observed that the electromagnetic torque oscillation is stable during transient state and settles within 2 seconds after disturbance. Also, rotor speed deviation due to transient state at 70% of K is stable and gets damped in 5 seconds, as shown in Figure 10.15, even without implementing controllers.

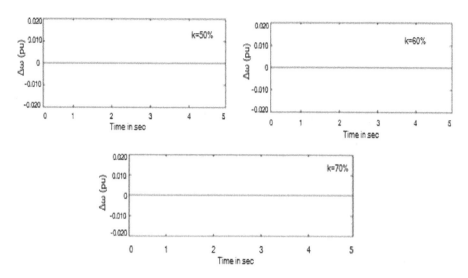

FIGURE 10.13   Rotor speed deviation ($\Delta\omega$) at 100-MW wind turbine generator for $K = 50\%$, 60%, and 70% during steady state.

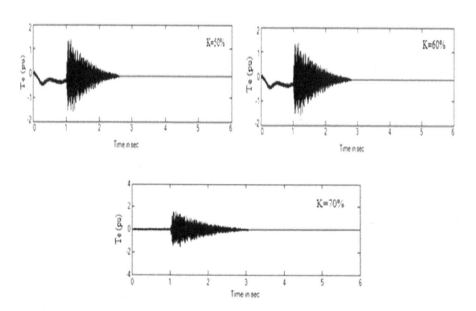

FIGURE 10.14   Electromagnetic torque (T_e) at 100-MW wind turbine generator for $K = 50\%$, 60%, and 70% during transient state.

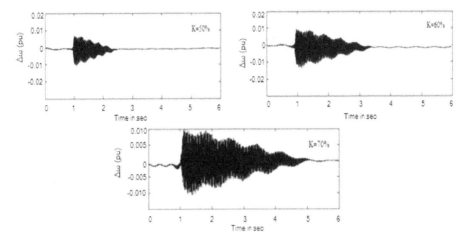

FIGURE 10.15 Rotor speed deviation ($\Delta\omega$) at 100-MW wind farm for $k = 50\%$, 60%, and 70% during transient state.

10.6.3 STEADY-STATE SSR ANALYSIS IN STEAM TURBINE GENERATOR WITHOUT CONTROLLERS

The eigenvalue analysis proving the occurrence of SSR oscillation effect is validated with the time domain simulation results. The time domain simulations are performed for 50%, 60%, 70%, and 80% series compensation levels K. According to Figure 10.16, the torque between generator and low-pressure turbine (T(Gen-LP)) pu shows the presence of SSR oscillation at the values of 70% and 80%, the value of torque attains a maximum of 20 pu at 40 seconds and 28 pu at 40 seconds, respectively. However, there is no oscillation occurring at 50% and 60% values of K. From Figure 10.17, the rotor speed deviation ($\Delta\omega$) tends to originate at 70% and 80% values of K and reaches the maximum values of 0.22 and 0.26 pu at $t = 40$ seconds.

10.6.4 STEADY-STATE SSR ANALYSIS IN STEAM TURBINE GENERATOR WITH CONTROLLERS

The previous study indicates that during steady-state conditions, the SSR oscillations are observed at 70% and 80% series compensation levels K. Hence, the system is tested with fuzzy logic controlled SSSC and TCSC and analyzed in damping SSR oscillations in a modified IEEE SBM system.

CASE I TORQUE BETWEEN GENERATOR AND LOW-PRESSURE TURBINE T(GEN-LP)PU DURING STEADY STATE WITH FUZZY-CONTROLLED SSSC AND TCSC IN A MODIFIED IEEE SBM SYSTEM

Figure 10.18 shows the T(Gen-LP)pu during steady state at 70% and 80% series compensation with fuzzy logic controlled SSSC. Based on Figure 10.18, with SSSC

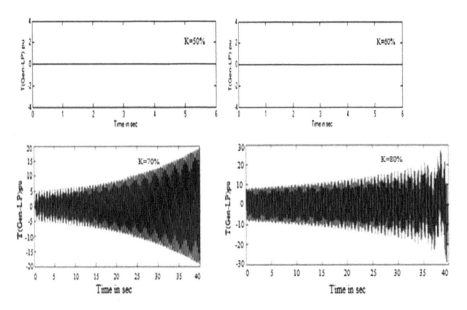

FIGURE 10.16 *T*(Gen-LP)pu during steady state at 50%, 60%, 70%, and 80% series compensation without controller.

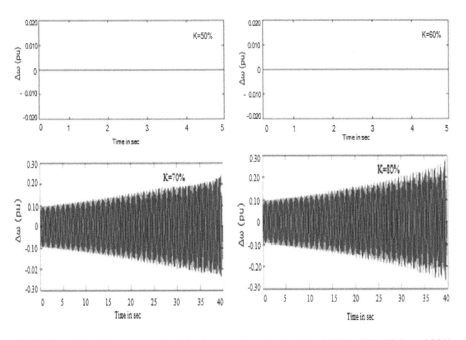

FIGURE 10.17 Rotor speed deviation (Δω) during steady state at 50%, 60%, 70%, and 80% series compensation without controller.

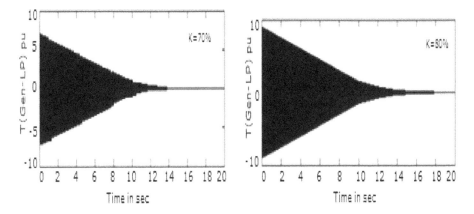

FIGURE 10.18 T(Gen-LP)pu during steady state at 70% and 80% series compensation with fuzzy logic controlled SSSC.

controller at $K = 70\%$, the Gen-LP torque oscillation attains a maximum value of 7.5 pu and due to the effect of controller the oscillation gets damped at 14 seconds. At $K = 80\%$, Gen-LP torque oscillation reaches a maximum value magnitude of 9 pu and the oscillation gets damped at $t = 18$ seconds.

Figure 10.19 shows the T(Gen-LP)pu during steady state at 70% and 80% series compensation with fuzzy logic controlled TCSC. According to Figure 10.19, the Gen-LP torque oscillation at $K = 70\%$ reaches a maximum value of 9 pu, and with the aid of TCSC controller the oscillation gets completely damped at 18 seconds. At $K = 80\%$, Gen-LP torque oscillation takes 21 seconds to get completely damped from a magnitude of 11 pu.

CASE II ROTOR SPEED DEVIATION($\Delta\Omega$) DURING STEADY STATE WITH FUZZY-CONTROLLED SSSC AND TCSC IN A MODIFIED IEEE SBM SYSTEM

The rotor speed deviation ($\Delta\omega$) is another important parameter to be observed. Figure 10.20 shows rotor speed deviation ($\Delta\omega$) during steady state at 70%

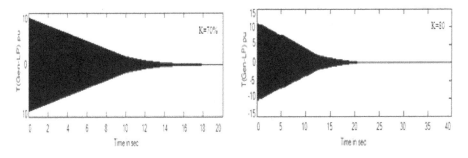

FIGURE 10.19 T(Gen-LP)pu during steady state at 70% and 80% series compensation with fuzzy logic controlled TCSC.

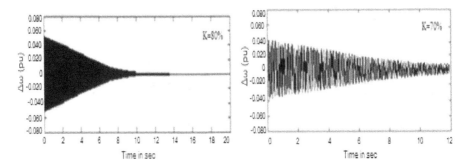

FIGURE 10.20   Rotor speed deviation ($\Delta\omega$) during steady state at 70% and 80% series compensation with fuzzy logic controlled SSSC.

and 80% series compensation with fuzzy logic controlled SSSC. From the figure, it is observed that the rotor speed oscillates and tends to reach a maximum value of 0.04 pu at $K = 70\%$ and 0.05 pu at $K = 80\%$. Due to the impact of SSSC controller, the oscillation gets damped at 12 and 13 seconds, respectively.

It can be observed from Figure 10.21 that rotor speed deviation ($\Delta\omega$) reaches a maximum value of 0.08 pu at $K = 70\%$, and 0.10 pu at $K = 80\%$. However, due to the impact of TCSC controller, the oscillation gets damped at 17.5 and 18 seconds, respectively.

10.6.5   Transient State SSR Effect Analysis in Steam Turbine Generator without Controllers

For analysis of transient state, a three-phase fault of 0.05 seconds duration is initiated at 1 second in the series-compensated line study system. Figure 10.22 shows the impact of the fault on the Gen-LP torque oscillation at various levels of series compensation K. The value of K is set to 50%, 60%, 70%, and 80%. Whenever the value of series compensation increases from 50% to 80%, the oscillation at the G-LP shaft section also increases. At $k = 50\%$ and 60% series compensation, it is observed that the Gen-LP

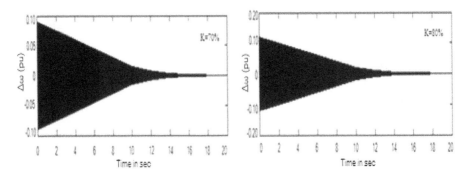

FIGURE 10.21   Rotor speed deviation ($\Delta\omega$) during steady state at 70% and 80% series compensation with fuzzy logic controlled TCSC.

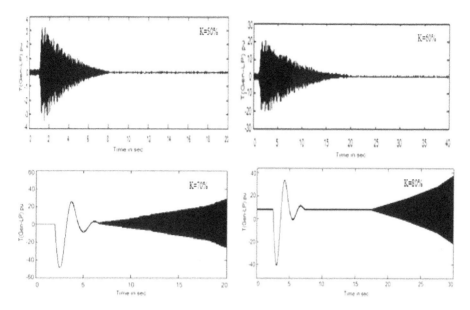

FIGURE 10.22 *T*(Gen-LP)pu during transient state at 50%, 60%, 70%, and 80% series compensation without controller.

shaft oscillation attains the maximum values of 3.2 and 20 pu and gets damped at $t = 8$ seconds and $t = 18$ seconds, respectively. However, for other sets of K values, the oscillations get increased, as depicted in Figure 10.22. A similar fault study is carried out on a modified IEEE SBM system, as shown in Figure 10.7. Figure 10.23 shows the impact of the fault on the rotor speed deviation ($\Delta\omega$) at different levels of series compensation K. At 50% and 60% series compensation level, ($\Delta\omega$) attains a maximum value of 0.009 and 0.018 pu, respectively, and reaches steady state from there on. For the remaining values of K at 70% and 80%, the rotor speed deviation gets increased and goes uncontrollable, as depicted in Figures 10.22 and 10.23.

10.6.6 TRANSIENT-STATE SSR ANALYSIS IN STEAM TURBINE GENERATOR WITH CONTROLLERS

In order to study the performance of fuzzy logic controlled SSSC and TCSC on SSR oscillations effects during transient state, simulation runs were done with different levels of series compensation percentage K.

CASE I TORQUE BETWEEN GENERATOR AND LOW-PRESSURE TURBINE *T*(GEN-LP)PU DURING TRANSIENT STATE WITH FUZZY-CONTROLLED SSSC AND TCSC IN A MODIFIED IEEE SBM SYSTEM

Figure 10.24 shows the *T*(Gen-LP)pu during transient state at 50%,60%,70%, and 80% series compensation with fuzzy logic controlled SSSC. It can be observed that

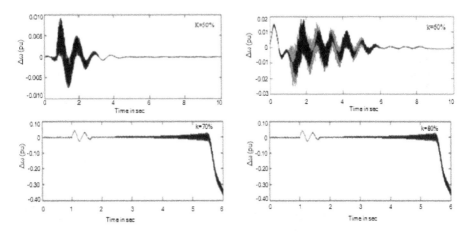

FIGURE 10.23 Rotor speed deviation ($\Delta\omega$) during transient state at 50%, 60%, 70%, and 80% series compensation without controller.

due to the influence of controllers, the damping of turbine-generator oscillations is achieved at all levels of series compensation. But one important observation made is that as K value increases from 50% to 80%, turbine-generator oscillatory torque magnitude also increases, thereby increasing the settling time of oscillatory torque.

Figure 10.25 shows the T(Gen-LP)pu during transient state at 50%, 60%, 70%, and 80% series compensation with fuzzy logic controlled TCSC. At 50% of series compensation, it is observed that the value of Gen-LP torque reaches a maximum of 1.5 pu and gets completely damped at $t = 5$ seconds.

For maximum value of series compensation $K = 80\%$, the value of Gen-LP torque attains 10 pu and settles at 24 seconds. At 60% and 70% compensation, the

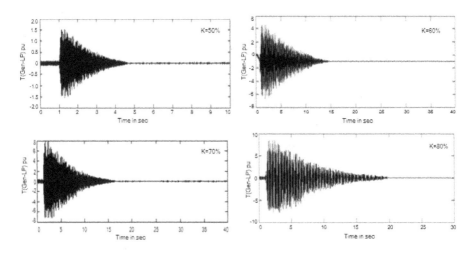

FIGURE 10.24 T(Gen-LP)pu during transient state at 50%, 60%, 70%, and 80% series compensation with fuzzy logic controlled SSSC.

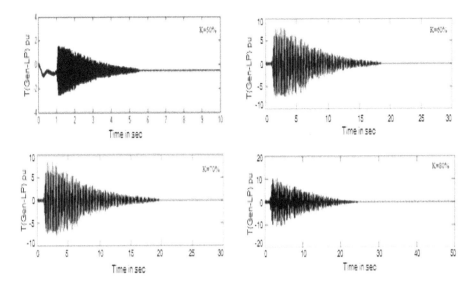

FIGURE 10.25   *T*(Gen-LP)pu during transient state at 50%, 60%, 70%, and 80% series compensation with fuzzy logic controlled TCSC.

magnitudes of oscillatory torques are 7.5 and 8 pu, and it took 18 and 19 seconds for complete damping of oscillation.

CASE II Rotor Speed Deviation ($\Delta\Omega$) during Transient State with Fuzzy-Controlled SSSC and TCSC in an IEEE SBM System

Figure 10.26 shows the influence of fuzzy controller in SSSC in damping of rotor speed deviation in IEEE SBM system. There exists proportionality between the series compensation percentage (*K*) and settling time. With increase in *K*, the rotor speed deviation increases, which is successfully damped by SSSC-based fuzzy controller.

The influence of fuzzy-controlled TCSC in damping of rotor speed deviation when the system is subjected to different levels of series compensation is shown in Figure 10.27. For minimum series compensation value of 50%, the rotor speed deviation takes 7 seconds to completely settle down from a magnitude of 0.0075 pu. For *K* = 60%, the value of rotor speed deviation is 0.020 pu and it takes 11 seconds to get completely damped. In case of 70% and 80% values of *K*, the settling time of rotor speed oscillations are 20 and 27 seconds from the peak values of 0.070 and 0.090 pu rotor speed deviation, respectively.

All the time domain simulation results are numerically tabulated in Tables 10.1–10.4.

10.6.7   Case study under Different Loaded Condition

To further test the performance of the proposed controller under different loaded conditions, the modified IEEE SBM system is subjected to transient-state analysis

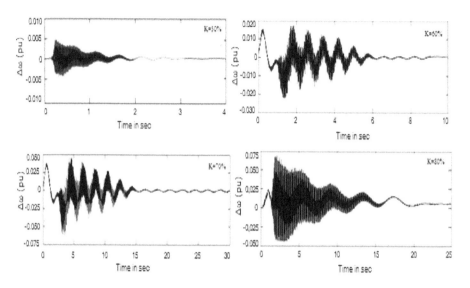

FIGURE 10.26   Rotor speed deviation ($\Delta\omega$) during transient state at 50%, 60%,70%, and 80% series compensation with fuzzy logic controlled SSSC.

under 70% and 80% values of series compensation. The impact of fuzzy-controlled TCSC in damping of rotor speed deviation when the system is subjected to different levels of series compensation is shown in Figure 10.28. For series compensation value of 70%, the rotor speed deviation takes 16 seconds to completely settle down from a magnitude of 0.055 pu. For $K = 80\%$, the magnitude of rotor speed deviation is 0.075 pu and it takes 21 seconds to get completely damped.

In case of a fuzzy-controlled SSSC in a modified IEEE SBM system, for 70% and 80% values of K, the settling time of rotor speed oscillations are 14 and 16 seconds, respectively, which is shown in Figure 10.29.

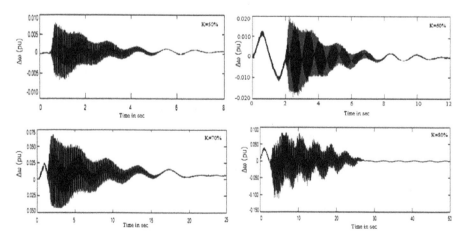

FIGURE 10.27   Rotor speed deviation ($\Delta\omega$) during transient state at 50%, 60%, 70%, and 80% series compensation with fuzzy logic controlled TCSC.

TABLE 10.1

T(Gen-LP)pu in a Modified IEEE SBM System at Different Levels of Series Compensation (*K*), with and without Controllers during Steady State

Results	Without Controller		With SSSC		With TCSC	
	Magnitude		Magnitude		Magnitude	
K (%)	(pu)	Time (s)	(pu)	Time (s)	(pu)	Time (s)
50 and 60	0	0	–	–	–	–
70	20.0	40.0	7.5	14.0	9.0	18.0
80	28.0	40.0	9.0	18.0	11.0	21.0

TABLE 10.2

Rotor Speed Deviation ($\Delta\omega$)pu in a Modified IEEE SBM System at Different Levels of Series Compensation (*K*), with and without Controllers during Steady State

Results	Without Controller		With SSSC		With TCSC	
	Magnitude		Magnitude		Magnitude	
K (%)	(pu)	Time (s)	(pu)	Time (s)	(pu)	Time (s)
50	0	0	0	0	0	0
60	0	0	0	0	0	0
70	0.22	40.0	0.04	12.0	0.08	17.5
80	0.26	40.0	0.05	13.0	0.10	18.0

TABLE 10.3

T(Gen-LP)pu in a Modified IEEE SBM Systems at Different Levels of Series Compensation (*K*), with and without Controllers during Transient State

Results	Without Controller		With SSSC		With TCSC	
	Magnitude		Magnitude		Magnitude	
K (%)	(pu)	Time (s)	(pu)	Time (s)	(pu)	Time (s)
50	3.2	8.0	1.5	4.5	1.5	5.5
60	20.0	18.0	5.0	14.0	7.5	18.0
70	30.0	20.0	8.0	16.0	8.0	19.0
80	38.0	30.0	9.0	20.0	10.0	24.0

TABLE 10.4

Rotor Speed Deviation ($\Delta\omega$)pu in a Modified IEEE SBM System at Different Levels of Series Compensation (K), with and without Controllers during Transient State

	Without Controller		With SSSC		With TCSC	
Results	Magnitude		Magnitude		Magnitude	
K (%)	(pu)	Time (s)	(pu)	Time (s)	(pu)	Time (s)
50	0.009	5.5	0.005	3.2	0.0075	5.0
60	0.020	11.0	0.018	8.5	0.020	10.0
70	−0.400	6.0	0.050	17.0	0.070	20.0
80	−0.400	6.0	0.072	21.0	0.090	27.0

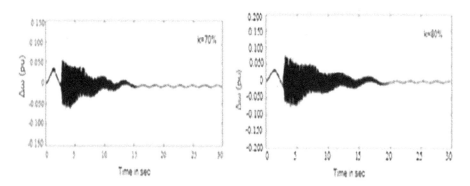

FIGURE 10.28   Rotor speed deviation ($\Delta\omega$) during transient state at 70% and 80% series compensation with fuzzy logic controlled TCSC.

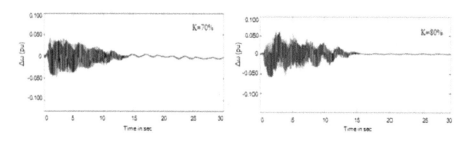

FIGURE 10.29   Rotor speed deviation ($\Delta\omega$) during transient state at 70% and 80% series compensation with fuzzy logic controlled SSSC.

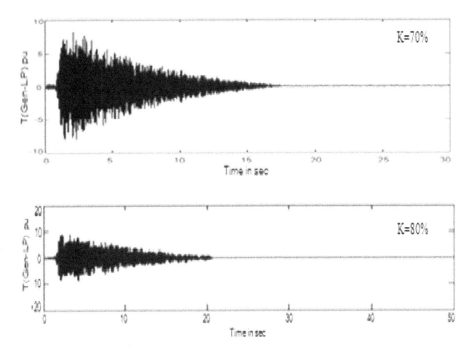

FIGURE 10.30 *T*(Gen-LP)pu during transient state at 70% and 80% series compensation with fuzzy logic controlled TCSC.

The impact of fuzzy-controlled TCSC in damping torque between generator and low-pressure turbine in the modified IEEE SBM system at different levels of series compensation and load is shown in Figure 10.30. For series compensation value of 70%, the magnitude of *T*(Gen-LP) takes 17 seconds to completely settle down from a magnitude of 8 pu. For *K* = 80%, the magnitude of *T*(Gen-LP) is 9 pu and it takes 21 seconds to get completely damped.

In case of a fuzzy controlled SSSC in a modified IEEE SBM system, for 70% and 80% values of *K*, the settling time for *T*(Gen-LP) are 13 and 18 seconds, respectively, as shown in Figure 10.31.

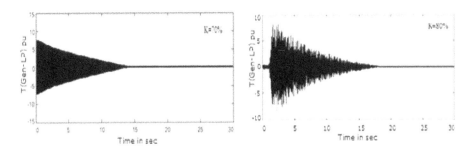

FIGURE 10.31 *T*(Gen-LP)pu during transient state at 70% and 80% series compensation with fuzzy logic controlled SSSC.

10.7   CONCLUSION

This chapter clearly explains the time domain simulations carried out on the 100-MW wind-generating farm considered as a single unit, connected to an IEEE SBM system. The modified IEEE SBM system is tested at various series compensation levels K under various operating scenarios. According to simulation results, it is conveyed that at low level of series compensation, no SSR oscillations is found in low-power wind farms when connected to an IEEE SSR SBM system during steady state. However, with 60% and 70% series compensation, it is found that SSR starts to occur and consequently the SSR oscillation damps quickly with the help of fuzzy logic based SSSC and TCSC controllers. Based on the simulation results, it is concluded that lower size wind farms are not prone to SSR oscillations effect. Simulation results and tabulated data of generator and low-pressure turbine and rotor speed deviation in a modified IEEE SBM system unit prove the superior oscillation damping effectiveness of SSSC when compared with TCSC during both steady-state and transient-state conditions. The new modified IEEE SBM system is also subjected to SSR oscillation damping study with FLC SSSC and FLC TCSC with different operating load conditions. From the results it is observed that FLC SSSC can damp SSR oscillation at a faster rate even under different loaded conditions.

10.8   SUMMARY

   i. A wind power integrated IEEE SBM system is tested at various series compensation levels under both steady-state and transient-state conditions.
  ii. Based upon the simulation results shown in Figures 10.12–10.27, it is proved that at low levels of series compensation, there will be no SSR oscillations in low-power wind farms (100 MW) when connected to an IEEE SBM system during steady state.
 iii. For higher levels of series compensation($K = 60\%$ and 70%), it is observed that SSR starts to occur and thus the SSR oscillation damps quickly with the help of fuzzy logic controlled SSSC and TCSC.
  iv. Simulation result data of torque between generator and low-pressure turbine and rotor speed deviation in a modified IEEE SBM system data are tabulated and the tabulated data prove the superior oscillation damping effectiveness of SSSC when compared to TCSC during steady-state as well as transient-state conditions.
   v. Simulation results shown in Figures 10.28–10.31 in a modified IEEE SBM system under different load conditions also prove the superior effectiveness of SSSC over TCSC in SSR oscillations damping.

REFERENCES

Cheng, M & Zhu, Y, 2014,'The state of the art of wind energy conversion systems and technologies: a review', Energy Conversion and Management, vol. 88, pp. 332–347.

Chilwal, B &Mishra, PK, 2020,'A survey of fuzzy logic inference system and other computing techniques for agricultural diseases,' in: Singh Tomar, G, Chaudhari, N, Barbosa, J, and Aghwariya, M (Eds.) International Conference on Intelligent Computing and Smart Communication 2019, Algorithms for Intelligent Systems, Springer, Singapore. https://doi.org/10.1007/978-981-15-0633-8_1

Fan, YW and Miao, YF, 2012, 'Effect of electronic word-of-mouth on consumer purchase intention: The perspective of gender differences'. International Journal of Electronic Business Management, 10, 175–181.

Gahramani, H, Lak, A, Farsadi, M & Hosseini, H, 2013, 'Mitigation of SSR and LFO with a TCSC based-conventional damping controller optimized by the PSO algorithm and a fuzzy logic controller', Turkish Journal of Electrical Engineering & Computer Sciences, vol. 21, pp. 1302–1317.

Gao, D, Jin, Z & Lu, Q, 2008, 'Energy management strategy based on fuzzy logic for a fuel cell hybrid bus', Journal of Power Sources, vol. 185, no. 1, pp. 311–317.

Hosseini, SMH, Samadzadeh, H, Olamaei, J & Farasadi, M, 2013, 'SSR mitigation with SSSC thanks to fuzzy control' Turkish Journal of Electrical Engineering & Computer Sciences, vol. 21, pp. 2294–2306.

Jang, JSR, Sun, CT & Mizutani, E, 1997, Neuro-Fuzzy and Soft Computing, Prentice-Hall of India Publishers, New Delhi.

Pachauri, RK, Kumar, H, Gupta, A & Chauhan, YK, 2016, 'Pitch angle controlling of wind turbine system using proportional-integral/fuzzy logic controller', Proceedings of 3rd International Conference on Advanced Computing, Networking and Informatics, pp. 55–63.

Rahim, AHMA, 2004, 'A minimum-time based fuzzy logic dynamic braking resistor control for sub-synchronous resonance', International Journal of Electrical Power and Energy Systems, vol. 26, pp. 191–198.

Wang, L & Truong, DN, 2013,'Stability enhancement of a power system with a PMSG-based and a DFIG-based offshore wind farm using a SVC with an adaptive-network-based fuzzy inference system', IEEE Transactions on Industrial Electronics, vol. 60, no. 7, pp. 2799–2807.

11 Adaptive Neuro-Fuzzy Inference Strategy Controlled TCSC and SSSC in a Modified IEEE SBM System for Damping SSR Oscillations

LEARNING OUTCOME

i. To study about the IEEE SBM system with wind power integrated system.
ii. To design ANFIS controller for damping SSR oscillation.
iii. To analyze the effect of ANFIS-controlled SSSC and TCSC in damping SSR oscillation effects during steady state.
iv. To demonstrate the effect of ANFIS-controlled SSSC and TCSC in damping SSR oscillation effects during transient state.

11.1 INTRODUCTION

This chapter discusses the flexible alternating current transmission system (FACTS) devices, namely, thyristor-controlled series capacitor (TCSC) and static synchronous series compensator (SSSC), for effective damping of subsynchronous resonance (SSR) oscillations in series-compensated transmission network using adaptive neuro-fuzzy inference strategy (ANFIS).The addition of series compensation to existing power system network is well proved in various literature surveys discussed in Chapter 1. ANFIS is a proven common technique owing to its being computationally less expensive and transparent and producing results as robust as statistical models (Muljono et al. 2016; and Chang 2006; Kurnaz et al. 2010; Zounemat-Kermani and Teshnehlab 2008). Dehghani et al. (2019) proposed Grey Wolf Optimization technique for prediction of hydropower generation and proved that the results obtained from this unique Grey Wolf ANFIS systems are better than other fuzzy expert systems. Another major advantage is that ANFIS can be interpreted as local linearization model for model estimation, proving that it has a good applicability in system modeling. Optimization using ANFIS for Groundwater Potential Mapping (Termeh et al. 2019) was presented. This system is designed to allow if-then rules and membership functions (fuzzy logic) to be constructed based on the historical data, and the design also includes the adaptive nature for automatic tuning of the membership functions (Yan et al. 2010). This ANFIS controller utilizes the full advantages of both neural network controller and fuzzy logic controller so that effective damping of SSR-related oscillations evolving during different operating scenarios is achieved. (Ain et al. 2020), in his work, proposed a comparison between the effects of SSSC and TCSC tested under various fault conditions. This work is a noteworthy study for power system stability enhancement with SSSC and TCSC controllers. ANFIS-based controller performance in SSSC and TCSC at a modified IEEE Second Benchmark (SBM) system to reduce the subsynchronous oscillation is discussed in this chapter.

The IEEE SBM system with low-power wind generation integrated system is carried out using MATLAB software package. The fuzzy logic based controllers used in Chapter 10 develops a control signal, based on random rule base firing. Hence, the controller output is also random and effective results (fast settling time of oscillations) are to be improved. By proper selection of a suitable membership functions and in turn choosing the proper basis for the rules based on the situation, the ANFIS controller that is a cohesive approach of effective control will give excellent results. The damping time of SSR-related oscillations in the power system under steady-state and transient-state conditions, with and without ANFIS-controlled SSSC and TCSC, are tabulated and compared under various levels of series compensation. The SSR oscillations in the power system are improved by measuring system parameters like rotor speed deviation and line current, which are utilized as input parameters for SSSC and TCSC controllers, respectively. The ANFIS control strategy that provides excellent results statistically is the highlight of this chapter. The simulation results in time domain obtained using MATLAB/SIMULINK software presented at the end of this chapter prove that if the designed control is more effective, it will have faster settling time for SSR oscillations.

11.2   IEEE SBM SYSTEM WITH WIND POWER INTEGRATED SYSTEM

A standard IEEE SBM system consists of parallel transmission lines, four-mass steam generators connected to an infinite bus, of which one is series compensated. However, based on eigenvalue analysis and results obtained in previous literatures, this type of steam turbine system is prone to SSR-related oscillations owing to the impact of series compensation. With fuzzy-controlled series compensation using SSSC and TCSC discussed in the previous chapters, it was proven that with appropriate selection of controllers and FACTS devices, the SSR-related oscillations can be effectively damped.

The modified version of IEEE SBM comprises low-power DFIG wind farm combined with a two-stage four-mass steam turbine generator (TG), as shown in Figure 11.1. A more detailed technical aspect of a conventional four-mass steam turbine generator was explained in Chapter 10. The technical specifications of steam and wind generation units are given in Appendix in detail.

Figures 11.2 and 11.3 show the revised IEEE SBM system with SSSC and TCSC, respectively.

On the occurrence of three-phase fault in the system, a phenomenon exists that excites the oscillation on the turbine-generator shafts as well as amplifies their

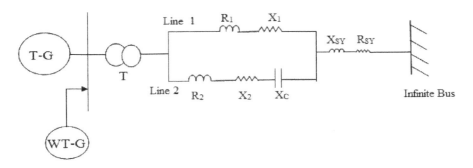

FIGURE 11.1   Modified IEEE SBM.

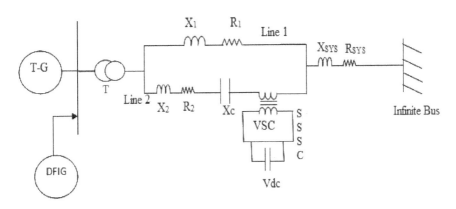

FIGURE 11.2   Modified IEEE SBM with SSSC.

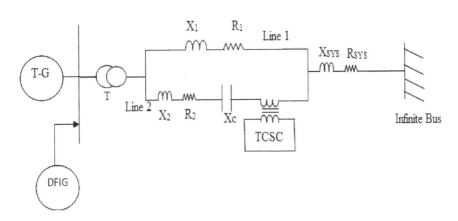

FIGURE 11.3   Modified IEEE SBM with TCSC.

torque. The main contribution of this study is to estimate the effect of series compensation individually in wind-generating unit and two-stage four-mass steam turbine-generators that was not performed so far and the impact of ANFIS-controlled SSSC and TCSC in damping the SSR oscillations.

11.3 REPRESENTATION OF SSSC AND TCSC WITH ANFIS CONTROLLER

The main functional blocks of ANFIS are as follows:

    i. A rule base containing a number of fuzzy if-then rules.
   ii. A database defining the membership functions of the fuzzy sets used in the fuzzy rules.
  iii. A decision-making unit that performs the inference operations on the rules.
  iv. A fuzzification interface converts the crisp inputs into degrees of match with linguistic values.
   v. A defuzzification interface converts the fuzzy results of the inference into a crisp output.

The rule base and the database are jointly termed as the knowledge base.

As discussed in Chapter 10, the fuzzy controller which is replaced by an ANFIS controller herein a modified IEEE SBM system. Hence, Figures 11.4 and 11.5 show the schematic representation of SSSC and TCSC with ANFIS controller.

11.4 DESIGN OF ANFIS CONTROLLER FOR DAMPING SSR OSCILLATION

Damping the SSR oscillation by designing an ANFIS-based control strategy in FACTS device is discussed in this section. The advantages of fuzzy logic and neural network control are used in framing the situation-dependent rule base in an ANFIS controller, which becomes an effective method of control. The schematic diagram of the ANFIS controller for SSSC and TCSC is displayed in Figure 11.6.

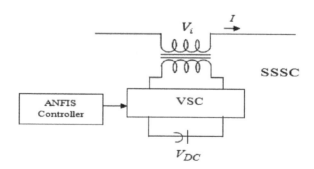

FIGURE 11.4   Basic scheme of SSSC with ANFIS controller.

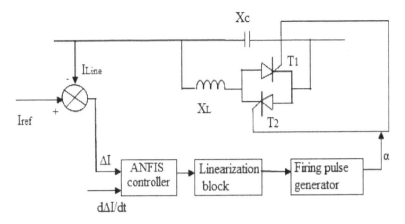

FIGURE 11.5   Basic scheme of TCSC with ANFIS controller.

According to Figure 11.6, the input signals to ANFIS controller are the error and the change in error, which is modeled by Equations (11.1) and (11.2):

$$e(k) = \omega_{ref} - \omega_r \tag{11.1}$$

$$\Delta e(k) = e(k) - e(k-1) \tag{11.2}$$

The ANFIS scheme proposed is of five layers, and with the help of neural network techniques, proper rule base is selected using the back-propagation algorithm. A neuro-fuzzy system will have a neural-network architecture framed through fuzzy reasoning. The structured knowledge is codified into fuzzy rules, whereas the adapting and learning capabilities of neural networks are retained. Sugeno-type fuzzy inference system (FIS) controller is used in the proposed ANFIS controller. The FIS controller parameters are decided by back-propagation technique (Zahra Amini & Abbas Kargar 2013). The rotor speed deviation ($\Delta\omega$) and its derivative ($d\,\Delta\omega/dt$) are taken as inputs for ANFIS SSSC-based damping controller, and the line current deviation (ΔI) and its derivative ($d\,\Delta I/dt$) as are taken as the input in TCSC-based ANFIS controllers. The output signal is calculated by using input variable

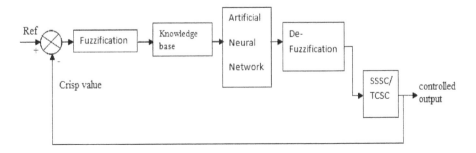

FIGURE 11.6   Block diagram of an ANFIS controller.

membership functions. For validating the effectiveness in SSSC and TCSC device, testing is carried out on a modified IEEE SBM system in the MATLAB/SIMULINK software.

The rule base of ANFIS controllers are as follows:

Example: A two inputs (x and y) and one output (z)

$$\textbf{Rule } 1 : \text{If} x \text{ is } A \text{ and } y \text{ is } B, \text{ then } f_1 = n_1 x + m_1 y + r_1 \tag{11.3}$$

$$\textbf{Rule } 2 : \text{If} x \text{ is } A \text{ and } y \text{ is } B, \text{ then } f_2 = n_2 x + m_2 y + r_2 \tag{11.4}$$

11.4.1 General Architecture of ANFIS

Layer 1: Every node i is an adaptive node with node function:

$$S_{1,i} = \mu_{A_i}(x), \text{ for } i = 1, 2$$

$$S_{1,i} = \mu_{B_{i-2}}(y), \text{ for } i = 3, 4$$

where x and y are the inputs of node i, and A_i (or B_i) is the linguistic label. Membership function for A (or B) can be any continuous and piecewise differentiable functions shown in (Figure 11.7).

The node functions in this layer,

$$\mu_A(x) = \exp\left[-\left(\frac{x - C_i}{a_i} \right)^2 \right] \tag{11.5}$$

where $\{c_i, a_i\}$ are the parameter set.

Layer 2: This layer is a fixed node with multiplication functionality:

$$S_{2,i} = w_i = \mu_{A_i}(x) X \mu_{B_i}(y) \tag{11.6}$$

where $i = 1, 2$, called firing strength of a rule.

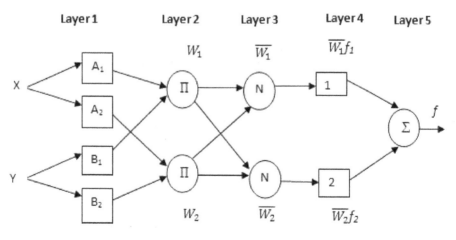

FIGURE 11.7   General ANFIS architecture.

Layer 3: Fixed node of normalization functionality, which performs the ratio computation of rule's firing strength to the summation of all rules' firing strengths:

$$S_{3,i} = \overline{w_i} = {w_i}\Big/{w_1} + w_2$$

where $i = 1,2$ is normalized firing strength.

Layer 4: Every node i of this layer is an adaptive node with a node function:

$$S_{4,i} = \overline{w_i} f_i = \overline{w_i}(n_i x + m_i y + r_i)$$

where $\overline{w_i}$ is the normalized firing strength from the third layer, and n_i, m_i, and r_i are the parameter set of the node.

Layer 5: This single layer node is a fixed node labeled (Σ), which calculates the overall output as the total of all incoming signals:

$$S_{5,i} = \sum_i \overline{w_i} f_i = \sum_i w_i f_i \Big/ \sum_i w_i$$

11.4.2 The Steps to be Carried Out for Constructing ANFIS-Controlled SSSC and TCSC

The following steps need to be carried out for constructing ANFIS-controlled SSSC and TCSC:

1. Designing of Simulink model with rule base simulated in fuzzy logic controller.
2. Gathering of trained data by ANFIS controller while simulating the model with fuzzy controller.
3. Training of data is provided by two inputs– speed deviation $\Delta\omega$ and its derivative $\dfrac{d\Delta\omega}{dt}$ in case of SSSC and ΔI and $\dfrac{\Delta I}{dt}$ for TCSC.
4. Using anfisedit generate ANFIS.fis file.
5. Using data obtained from step 2, generate the FIS with bell MFs.
6. The collected data are then trained using the generated FIS around 500 epochs in the considered application.
7. The trained ANFIS controller is then saved in Simulink model with SSSC and TCSC controllers.

Each training pattern comprises a set of input data and corresponding output data. The inputs for ANFIS in the case of SSSC will be the speed deviation and its derivative, and the input for ANFIS in case of TCSC will be the change in current and

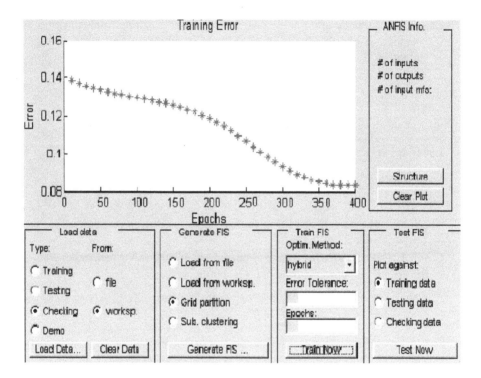

FIGURE 11.8   Training and updating of error.

its derivative. In this method, during forward pass, a least squares error (LSE) algorithm is used, and in the backward pass, a gradient descent algorithm (here back-propagation) is used to determine the optimal solution. Figure 11.8 shows the real-time training of error with epochs in case of SSSC.

The rule base for selection of proper rules using back-propagation algorithm is displayed in Table 11.1.

TABLE 11.1
Rule Base for Controlling the Firing Angle in SSSC and TCSC

$\Delta_e \Downarrow \Rightarrow e$	Nb	nm	ns	Ze	Ps	pm	pb
Nb	Nb	nb	nb	Nb	nm	ns	ze
Nm	Nb	nb	nm	Nm	ns	ze	ps
Ns	Nb	nm	ns	Ns	ze	ps	pm
Ze	Nb	nm	ns	Ze	ps	pm	pb
Ps	Nm	ns	ze	Ps	ps	pm	pb
Pm	Ns	ze	ps	Pm	pm	pb	pb
Pb	Ze	ps	pm	Pb	pb	pb	pb

The rule base explained in Table 9.1 in Chapter 9, used for Mamdani-based FLC for decision-making purposes, is also applied in this chapter for the decision-making purposes to design the ANFIS controller.

From the fuzzified variables in Table 11.1, the control decisions are made. The fuzzification process converts the crisp data into linguistic variables, which are inputs to the rule-based block; 49 rules sets are framed based on previous gained knowledge about the required applications. The rule-based block is connected to the neural network block. In order to provide effective creation of control signal, training of data is an important step in the selection of the proper rule base. When the proper rules are selected and fired, the necessary control signal required to obtain the optimal outputs are generated. The controlled output for the FACTS device, i.e., the firing angle for SSSC and TCSC, is the weighted average of the designed rule-based output obtained by the back-propagation algorithm.

11.5   CASE STUDY

The introduction of distributed generation has consequences, not only on the distribution network but also on the transmission grid as well as rest of the generators. The utility engineers' responsibility is to analyze the consequences by inserting wind generators into the power system, and in order to analyze the scenarios, an IEEE-modified SBM system is developed and SSR analysis with different levels of series compensation is done during steady-state and transient-state conditions using MATLAB software. A three-phase to ground fault at series-compensated line is initiated for testing the transient stability performance of SSSC and TCSC with ANFIS controller. The system under study is modeled using the inbuilt block sets given in the MATLAB software package. The simulations were carried out and from the time domain analysis, the potential for SSR at wind farm as well as at four-mass steam turbine generators are determined. The time domain simulations are carried out in two parts: (i) steady-state and (ii) transient-state conditions using MATLAB/SIMULINK.

11.5.1   STEADY-STATE SSR ANALYSIS IN STEAM TURBINE
GENERATOR WITHOUT CONTROLLERS

The time domain simulations using MATLAB are carried out for 50%, 60%, 70%, and 80% series compensation levels K. The results are depicted in Figure 11.9. From the figure, it is identified that there will be no torque oscillations at $K = 50\%$ and $K = 60\%$, whereas at series compensation level $K = 70\%$ and 80%, the value of torque between generator and low-pressure turbine attains a maximum of 20 pu at 40 seconds and 28 pu at 40 seconds, respectively.

Figure 11.10 shows the time domain simulations of rotor speed deviation ($\Delta\omega$) during steady state at 50%, 60%, 70%, and 80% series compensation without controller. With respect to Figure11.10, the series compensator at $K = 70\%$ and 80%, SSR oscillations exist at the values of 0.22 and 0.28 pu at 40 seconds, respectively.

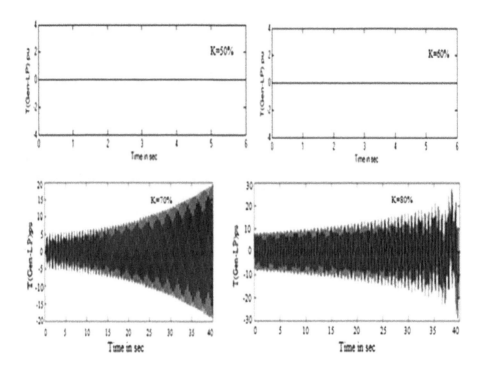

FIGURE 11.9 T(Gen-LP)pu during steady state at 50%, 60%, 70%, and 80% series compensation without controller.

11.5.2 STEADY-STATE SSR ANALYSIS IN STEAM TURBINE GENERATOR WITH CONTROLLERS

The existence of SSR oscillations during steady-state condition is observed at 70% and 80% series compensation levels K. However, the test system is tested with ANFIS-controlled SSSC and TCSC to reduce the oscillations.

CASE I TORQUE BETWEEN GENERATOR AND LOW-PRESSURE TURBINE T(GEN-LP)PU DURING STEADY STATE WITH ANFIS-CONTROLLED SSSC AND TCSC IN A MODIFIED IEEE SBM SYSTEM

Figure 11.11 shows the T(Gen-LP)pu during steady state at 70% and 80% series compensation with an ANFIS-controlled SSSC. From the figure, it is observed that with a SSSC controller, the Gen-LP torque oscillation at $K = 70\%$ reaches a maximum value of 6.2 pu, and owing to the effect of ANFIS controller, the oscillation gets damped at $t = 11$ seconds. Similarly, at $K = 80\%$, Gen-LP torque oscillation attains a maximum value of 10 pu and ANFIS controller settles the oscillation at $t = 14$ seconds.

Figure 11.12 shows the T(Gen-LP)pu during steady state at 70% and 80% series compensation with an ANFIS-controlled TCSC. From the figure, it is noted that the Gen-LP torque oscillation at $K = 70\%$ reaches a maximum value of 10 pu and due to

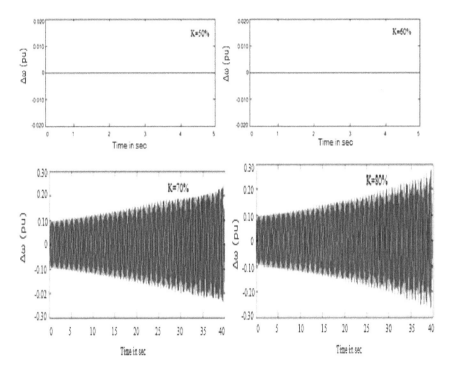

FIGURE 11.10   Rotor speed deviation ($\Delta\omega$) during steady state at 50%, 60%,70%, and 80% series compensation without controller.

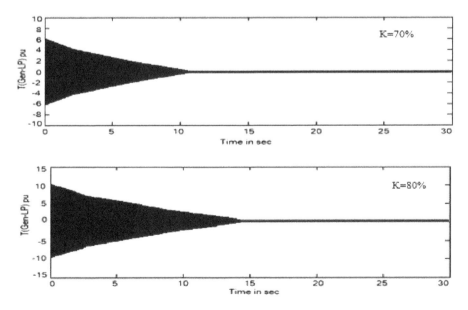

FIGURE 11.11 T(Gen-LP)pu during steady state at 70% and 80% series compensation with an ANFIS-controlled SSSC.

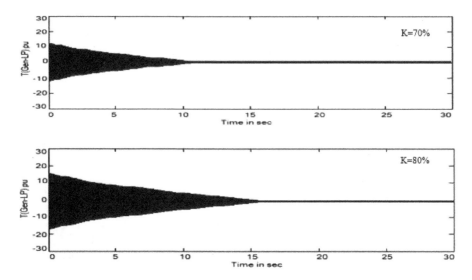

FIGURE 11.12 T(Gen-LP)pu during steady state at 70% and 80% series compensation with an ANFIS-controlled TCSC.

the effect of TCSC controller, the oscillation gets completely damped at 11 seconds. At $K = 80\%$, the Gen-LP torque oscillation reaches a maximum magnitude of 13 pu, and it takes 16 seconds to get completely damped from SSR oscillations.

CASE II Rotor Speed Deviation $(\Delta\Omega)$pu during Steady State with ANFIS-Controlled SSSC and TCSC in a Modified IEEE SBM System

The rotor speed deviation $(\Delta\omega)$ observed from Figure 11.13 tends to reach the maximum values of 0.040 pu at $K = 70\%$ and 0.070 pu at $K = 80\%$. Due to the impact of SSSC controller, the oscillation gets damped at 12 and 12 seconds, respectively.

Based on Figure 11.14, rotor speed deviation $(\Delta\omega)$ reaches the maximum values of 0.070 pu at $K = 70\%$ and 0.090 pu at $K = 80\%$. Due to the impact of TCSC controller, the oscillation gets damped at 12 and 12 seconds, respectively.

11.5.3   Transient-State SSR Effect Analysis in Steam Turbine Generator without Controllers

A three-phase fault of 0.05 seconds duration is simulated at the series-compensated line of the modified IEEE SBM system. Figure 11.15 shows the impact of the fault on the Gen-LP torque oscillation at different levels of series compensation K. The value of K is set to 50%, 60%, 70%, and 80%. As series compensation increased from 50% to 80%, the oscillation at the Gen-LP shaft section also increases.

At 50% and 60% series compensation, it is observed that the Gen-LP shaft oscillation reaches the maximum values of 3.2 and 20 pu and gets damped at 8 and 18 seconds, respectively, and for all other remaining values of K, the oscillations gets increased, as depicted in Figure 11.15.

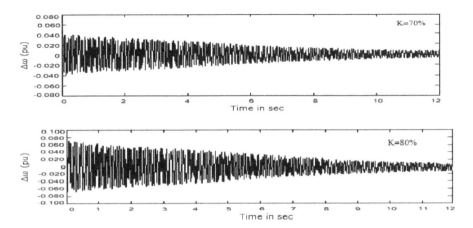

FIGURE 11.13 Rotor speed deviation ($\Delta\omega$)pu during steady state at 70% and 80% series compensation with an ANFIS-controlled SSSC.

A similar fault study is carried out on the modified IEEE SBM system, as shown in Figure 11.16. Figure 11.16 shows the impact of the fault on the rotor speed deviation ($\Delta\omega$) at different levels of series compensation K. The value of K is set to 50%, 60%, 70%, and 80%. At 50% and 60% series compensation level, ($\Delta\omega$) reaches the maximum values of 0.009 and 0.018 pu, respectively, and attains steady state from there on. For the remaining values of K at 70% and 80%, the rotor speed deviation gets increased uncontrollably, as depicted in Figure 11.16.

11.5.4 TRANSIENT-STATE SSR EFFECT ANALYSIS IN STEAM TURBINE GENERATOR WITH CONTROLLERS

For validation of an ANFIS controller in SSSC and TCSC in damping SSR oscillations effects during transient state, simulation runs were done with different levels of series compensation percent K.

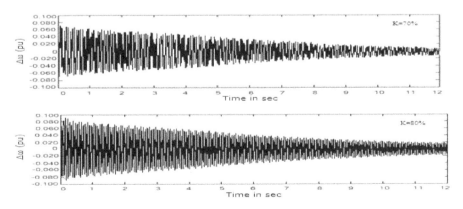

FIGURE 11.14 Rotor speed deviation ($\Delta\omega$) during steady state at 70% and 80% series compensation with an ANFIS-controlled TCSC.

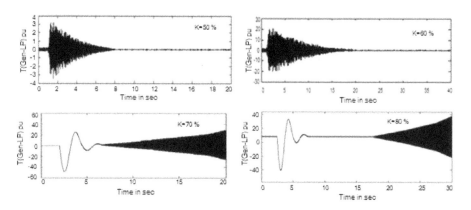

FIGURE 11.15 T(Gen-LP)pu during transient state at 50%,60%,70%, and 80% series compensation without controller.

CASE I Torque between Generator and Low-Pressure turbine T(Gen-LP)pu during Transient State with ANFIS-Controlled SSSC and TCSC in a Modified IEEE SBM System

The plots of torque between generator and low-pressure turbine for different value of series compensation with respect to settling time are depicted in Figure 11.17. All the plots are simulated with MATLAB software for a modified IEEE SBM system with an ANFIS-controlled SSSC. It can be observed that due to the influence of proposed controllers, the damping of turbine-generator oscillations is achieved at all levels of series compensation. But one important observation made is that as K value increases from 50% to 80%, turbine-generator oscillatory torque magnitude also increases, thereby increasing the settling time of oscillatory torque.

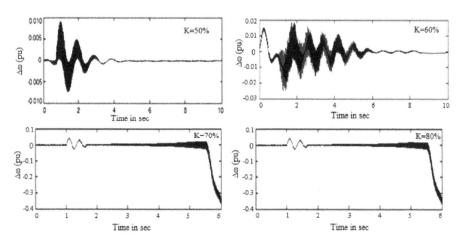

FIGURE 11.16   Rotor speed deviation ($\Delta\omega$) during transient state at 50%, 60%, 70%, and 80% series compensation without controller.

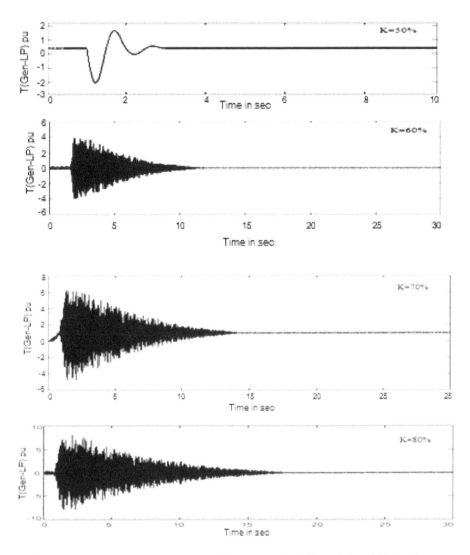

FIGURE 11.17 T(Gen-LP)pu during transient state at 50%, 60%, 70%, and 80% series compensation with an ANFIS-controlled SSSC.

The effect of ANFIS-controlled TCSC in damping of torque between generator and low-pressure turbine is shown in Figure 11.18. At 50% of series compensation, it could be observed that the value of Gen-LP torque reaches a maximum of 2.6 pu and gets damped completely at 3 seconds. For the maximum value of series compensation ($K = 80$%), the value of Gen-LP torque reaches 9.8 pu and settles at 21 seconds. At 60% and 70% compensation, the magnitudes of oscillatory torques are 4.5 and 8 pu, and it takes 13 and 16 seconds to get completely damped.

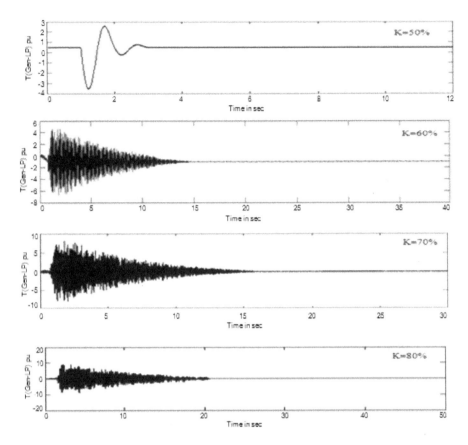

FIGURE 11.18 T(Gen-LP)pu during transient state at 50%, 60%, 70%, and 80% series compensation with an ANFIS-controlled TCSC.

CASE II Rotor Speed Deviation ($\Delta\Omega$) pu during Transient State with ANFIS-Controlled SSSC and TCSC in a Modified IEEE SBM System

Figure 11.19 shows the effect of ANFIS controller in SSSC for damping of rotor speed deviation. With increase in series compensation level, the rotor speed deviation in pu increases but the presence of ANFIS-controlled SSSC damps the SSR-related oscillations at a faster rate.

The impact of ANFIS-controlled TCSC in damping of rotor speed deviation when the system is subjected to different levels of series compensation is shown in Figure 11.20. For minimum series compensation value of 50%, the rotor speed deviation takes 5.2 seconds to completely settle down from a magnitude of 0.007 pu. For $K = 60\%$, the value of rotor speed deviation is 0.010 pu, and it takes 9 seconds to get completely damped. In case of 70% and 80% values of K, the settling time of rotor speed oscillations are 15 and 20 seconds, respectively.

From Tables 11.2–11.5, it is proved that the torque between generator and low-pressure turbine gets damped at a faster rate under the influence of ANFIS-controlled SSSC when

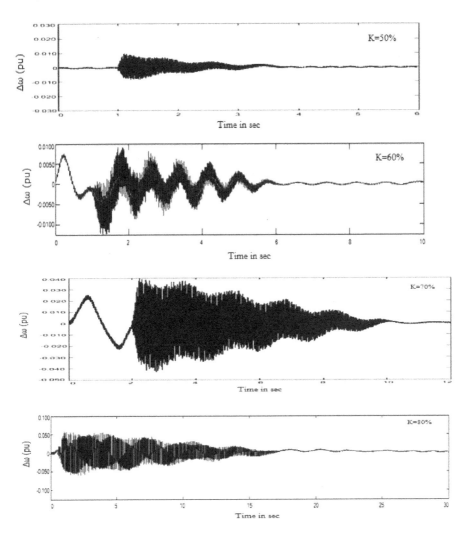

FIGURE 11.19 Rotor speed deviation ($\Delta\omega$) during transient state at 50%, 60%, 70%, and 80% series compensation with an ANFIS-controlled SSSC.

compared to TCSC-controlled compensation. The time taken to damp rotor speed deviation also reduces with both ANFIS-controlled SSSC and TCSC, but SSSC has superior performance in damping SSR oscillations when compared to TCSC in most cases.

CASE III CASE STUDY UNDER DIFFERENT LOADED CONDITIONS

The controller performance is studied under different loaded conditions and the modified IEEE SBM system is subjected to transient-state analysis under 70% and 80% values of series compensation. Figure 11.21 shows the rotor speed deviation ($\Delta\omega$) during transient state at 70% and 80%seriescompensation with ANFIS-controlled TCSC.

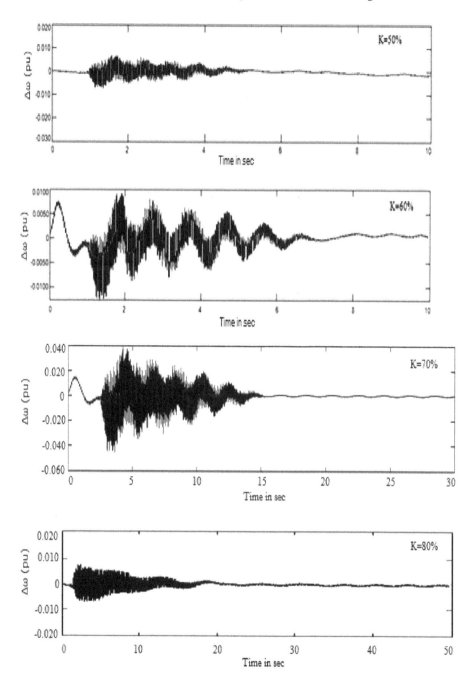

FIGURE 11.20   Rotor speed deviation ($\Delta\omega$) during transient state at 50%, 60%, 70%, and 80% series compensation with an ANFIS-controlled TCSC.

TABLE 11.2

T(Gen-LP)pu in a Modified IEEE SBM System at Different Levels of Series Compensation(K), with and without Controller during Steady State

	Without Controller		With SSSC		With TCSC	
Results K(%)	Magnitude (pu)	Time (s)	Magnitude (pu)	Time (s)	Magnitude (pu)	Time (s)
50	0	0	–	–	–	–
60	0	0	–	–	–	–
70	20.0	40.0	6.2	11.0	10.0	11.0
80	28.0	40.0	10.0	14.0	13.0	16.0

TABLE 11.3

Rotor Speed Deviation ($\Delta\omega$)pu during Steady State in a Modified IEEE SBM System at Different Levels of Series Compensation (K), with and without Controller

	Without Controller		With SSSC		With TCSC	
Results K (%)	Magnitude (pu)	Time (s)	Magnitude (pu)	Time (s)	Magnitude (pu)	Time (s)
50	0	0	0	0	0	0
60	0	0	0	0	0	0
70	0.22	40.0	0.040	12.0	0.070	12.0
80	0.28	40.0	0.070	12.0	0.090	12.0

TABLE 11.4

T(Gen-LP)pu in a Modified IEEE SBM System at Different Levels of Series Compensation (K), with and without Controllers during Transient State

	Without Controller		With SSSC		With TCSC	
Results K(%)	Magnitude (pu)	Time (s)	Magnitude (pu)	Time (s)	Magnitude (pu)	Time (s)
50	3.2	8.0	1.5	3.0	2.6	3.0
60	20.0	18.0	4.0	11.0	4.5	13.0
70	30.0	20.0	6.2	14.0	8.0	16.0
80	38.0	20.0	8.0	17.0	9.8	21.0

TABLE 11.5

Rotor Speed Deviation (Δω) in a Modified IEEE SBM System at Different Levels of Series Compensation (K), with and without Controllers during Transient State

	Without Controller		With SSSC		With TCSC	
Results	Magnitude	Time	Magnitude	Time	Magnitude	Time
K (%)	(pu)	(s)	(pu)	(s)	(pu)	(s)
50	0.009	5.0	0.009	3.8	0.007	5.0
60	0.018	7.0	0.010	7.0	0.010	9.0
70	−0.400	6.0	0.040	12.0	0.040	15.0
80	−0.400	6.0	0.050	17.0	0.070	20.0

For $K = 70\%$, the rotor speed deviation magnitude is 0.032 pu, and it takes 9 seconds to completely settle down. For $K = 80\%$, the magnitude of rotor speed deviation is 0.050 pu, and it takes 18 seconds to get completely damped.

Figure 11.22 shows the rotor speed deviation (Δω) during transient state at 70% and 80% series compensation with ANFIS-controlled SSSC based on Figure 11.22. It is found that for $K = 70\%$ and $K = 80\%$, the settling time of rotor speed deviation are 6 and 12 seconds, respectively.

Figure 11.23 shows the torque characteristics(Gen-LP)pu during transient state at 70% and 80% series compensation with ANFIS-controlled TCSC in modified IEEE SBM system. For $K = 70\%$, the magnitude of T(Gen-LP) oscillations reaches to 5 pu, and it takes 14 seconds to get completely damped. Similarly, from the figure, it is observed that for $K = 80\%$, the magnitude of T(Gen-LP) oscillation reaches to 8 pu, and it takes 16 seconds to get completely settle down.

Figure 11.24 shows the torque characteristics(Gen-LP) during transient state at 70% and 80% series compensation with ANFIS-controlled SSSC. It is observed from Figure 11.24 that with ANFIS-controlled SSSC, the settling time for T(Gen-LP) are 8 and 14 seconds, respectively.

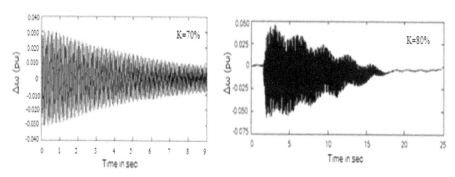

FIGURE 11.21   Rotor speed deviation (Δω) during transient state at 70% and 80% series compensation with ANFIS-controlled TCSC.

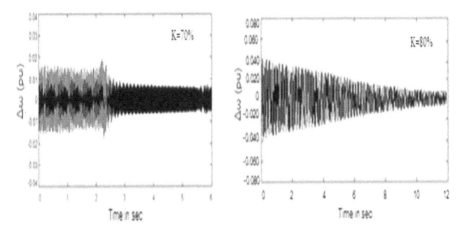

FIGURE 11.22 Rotor speed deviation ($\Delta\omega$) during transient state at 70% and 80% series compensation with ANFIS-controlled SSSC.

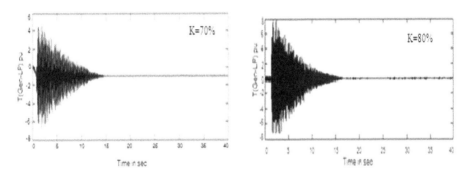

FIGURE 11.23 T(Gen-LP)pu during transient state at 70% and 80% series compensation with ANFIS-controlled TCSC.

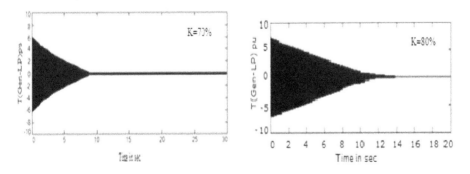

FIGURE 11.24 T(Gen-LP)pu during transient state at 70% and 80% series compensation with ANFIS-controlled SSSC.

11.6   CONCLUSION

This chapter clearly explains about the application of FACTS devices such as SSSC and TCSC using ANFIS controller to reduce the SSR oscillations in a modified IEEE SBM system. For effective validation, the simulation is carried out by MATLAB during steady state and transient state. The proposed method is achieved by injecting the voltage and varying line reactance by ANFIS techniques. The results obtained prove that both SSSC- and TCSC-based controllers have their individual control ability in damping SSR-related oscillation, whereas ANFIS-based SSSC has an edge over TCSC. A brief analysis was also done with modified IEEE SBM system subjected to SSR oscillation damping provided with ANFIS-controlled SSSC and TCSC under different loaded conditions. It is further proved that ANFIS-controlled SSSC reduces the SSR oscillation at a faster rate even under variable loaded conditions.

11.7   SUMMARY

i. The damping performance of SSSC and TCSC controllers using ANFIS technology is presented in this chapter.
ii. The modified IEEE SBM system is series compensated with ANFIS-controlled TCSC and SSSC.
iii. Simulation study and the obtained results illustrate the robustness of the ANFIS-based SSSC in damping SSR oscillation and rotor speed deviation when compared with ANFIS-based TCSC controllers under various series compensation levels (K) and operating conditions.
iv. An effective analysis is also carried out with variable load in an IEEE modified SBM system and the output results proved that ANFIS-controlled SSSC is better than the ANFIS-controlled TCSC.

REFERENCES

Ain, Q, Jamil, E, Hameed, S & Naqvi, KH, 2020, 'Effects of SSSC and TCSC for enhancement of power system stability under different fault disturbances', Australian Journal of Electrical and Electronics Engineering, vol. 17, no. 1, pp. 56–64.

Chang, F-J & Chang, Y-T, 2006, 'Adaptive neuro-fuzzy inference system for prediction of water level in reservoir', Advances in Water Resources, vol. 29, no. 1, pp.1-10.

Dehghani, M, Riahi-Madvar, H, Hooshyaripor, F, Mosavi, A, Shamshirband, S, Zavadskas EK & Chau, K-W, 2019, 'Prediction of hydropower generation using grey wolf optimization adaptive neuro-fuzzy inference system', Energies, vol. 12, p. 289.

Kurnaz, S, Cetin O & Kaynak, O, 2010, 'Adaptive neuro-fuzzy inference system based autonomous flight control of unmanned air vehicles', Expert Systems with Applications, vol.37, no.2, pp. 1229–1234.

Muljono, AB, Ginarsa IM & Nrartha, IMA, 2016, 'Stability enhancement of a large-scale power system using power system stabilizer based on adaptive neuro fuzzy inference system', International Journal of Electrical, Computer, Energetic, Electronic and Communication Engineering, vol. 10, no. 10, pp. 1334–1341.

Termeh, SVR, Khosravi K & Sartaj, M et al., 2019, 'Optimization of an adaptive neuro-fuzzy inference system for groundwater potential mapping', Hydrogeology Journal, vol. 27, pp. 2511–2534.

Yan, H, Zou Z & Wang, H, 2010, 'Adaptive neuro fuzzy inference system for classification of water quality status', Journal of Environmental Sciences, vol.22, no.12, pp. 1891–1896.

Zounemat-Kermani, M & Teshnehlab, M, 2008, 'Using adaptive neuro-fuzzy inference system for hydrological time series prediction', Applied Soft Computing, vol.8, no.2, pp. 928–936.

APPENDIX 1

TABLE A11.1
Network Impedances in per Unit Based on 100 MVA and 500 KVA Base for IEEE SSR SBM System

Parameter	Positive Sequence	Zero Sequence
R_T	0.0002	0.0002
X_T	0.0200	0.0200
R_1	0.0074	0.0220
X_{L1}	0.0800	0.2400
R_2	0.0067	0.0186
X_{L2}	0.0739	0.2100
R_{sys}	0.0014	0.0014
X_{SYS}	0.0300	0.0300

TABLE A11.2
Synchronous Machine Parameters at 600 MVA

Reactance	Value	Time Constants	
Ra	0.0045	T'_{do}	0.040 second
$Xd, X'd$, and $X''d$	1.650, 0.250, and 0.200	T''_{do}	0.090 second
$Xq, X'q$, and $X''q$	1.590, 0.460, and 0.200	T'_{qo}	0.040 second
		T'_{qo}	0.090 second

TABLE A11.3
IEEE SBM Rotor Model Data

Mass	Inertia lbm-ft²	Damping lbf-ft-second/rad	Shaft Section	Spring Constant lbf-ft/rad
EXC	1383	4.3	EXC-GEN	4.39×10^6
GEN	176204	547.9	LP-GEN	97.97×10^6
LP	310729	966.2	HP-LP	50.12×10^6
HP	9912	155.2	–	–

Index